Toxicologic Assessment of Jet-Propulsion Fuel 8

Subcommittee on Jet-Propulsion Fuel 8

Committee on Toxicology

Board on Environmental Studies and Toxicology

Division on Earth and Life Studies

NATIONAL RESEARCH COUNCIL
of the National Academies

THE NATIONAL ACADEMIES PRESS
Washington, D.C.
www.nap.edu

THE NATIONAL ACADEMIES PRESS 500 Fifth Street, N.W. Washington, D.C. 20001

NOTICE: The project that is the subject of this report was approved by the Governing Board of the National Research Council, whose members are drawn from the councils of the National Academy of Sciences, the National Academy of Engineering, and the Institute of Medicine. The members of the committee responsible for the report were chosen for their special competences and with regard for appropriate balance.

This project was supported by Contract No. DAMD17-99-C-9049 between the National Academy of Sciences and U.S. Department of Defense. Any opinions, findings, conclusions, or recommendations expressed in this publication are those of the author(s) and do not necessarily reflect the view of the organizations or agencies that provided support for this project.

International Standard Book Number 0-309-08715-5

Additional copies of this report are available from:

The National Academies Press
500 Fifth Street, NW
Box 285
Washington, DC 20055

800-624-6242
202-334-3313 (in the Washington metropolitan area)
http://www.nap.edu

Printed in the United States of America

THE NATIONAL ACADEMIES

Advisers to the Nation on Science, Engineering, and Medicine

The **National Academy of Sciences** is a private, nonprofit, self-perpetuating society of distinguished scholars engaged in scientific and engineering research, dedicated to the furtherance of science and technology and to their use for the general welfare. Upon the authority of the charter granted to it by the Congress in 1863, the Academy has a mandate that requires it to advise the federal government on scientific and technical matters. Dr. Bruce M. Alberts is president of the National Academy of Sciences.

The **National Academy of Engineering** was established in 1964, under the charter of the National Academy of Sciences, as a parallel organization of outstanding engineers. It is autonomous in its administration and in the selection of its members, sharing with the National Academy of Sciences the responsibility for advising the federal government. The National Academy of Engineering also sponsors engineering programs aimed at meeting national needs, encourages education and research, and recognizes the superior achievements of engineers. Dr. Wm. A. Wulf is president of the National Academy of Engineering.

The **Institute of Medicine** was established in 1970 by the National Academy of Sciences to secure the services of eminent members of appropriate professions in the examination of policy matters pertaining to the health of the public. The Institute acts under the responsibility given to the National Academy of Sciences by its congressional charter to be an adviser to the federal government and, upon its own initiative, to identify issues of medical care, research, and education. Dr. Harvey V. Fineberg is president of the Institute of Medicine.

The **National Research Council** was organized by the National Academy of Sciences in 1916 to associate the broad community of science and technology with the Academy's purposes of furthering knowledge and advising the federal government. Functioning in accordance with general policies determined by the Academy, the Council has become the principal operating agency of both the National Academy of Sciences and the National Academy of Engineering in providing services to the government, the public, and the scientific and engineering communities. The Council is administered jointly by both Academies and the Institute of Medicine. Dr. Bruce M. Alberts and Dr. Wm. A. Wulf are chair and vice chair, respectively, of the National Research Council

www.national-academies.org

SUBCOMMITTEE ON JET-PROPULSION FUEL 8

MELVIN E. ANDERSEN (*Chair*), Colorado State University, Fort Collins, Colorado; currently at CIIT-Centers for Health Research, Research Triangle Park, North Carolina
JANET M. BENSON, Lovelace Respiratory Research Institute, Albuquerque, New Mexico
SCOTT W. BURCHIEL, University of New Mexico, Albuquerque
RAKESH DIXIT, Merck Research Laboratory, West Point, Pennsylvania
ROBERT G. FELDMAN, Boston University School of Medicine, Boston, Massachusetts
SAM KACEW, University of Ottawa, Ottawa, Ontario, Canada
NANCY I. KERKVLIET, Oregon State University, Corvallis
MARTHA S. SANDY, California Environmental Protection Agency, Oakland
WILLIAM M. VALENTINE, Vanderbilt University Medical Center, Nashville, Tennessee

Staff

ABIGAIL E. MITCHELL, Project Director
NORMAN GROSSBLATT, Senior Editor
KELLY CLARK, Assistant Editor
MIRSADA KARALIC-LONCAREVIC, Research Assistant
JESSICA BROCK, Senior Project Assistant

COMMITTEE ON TOXICOLOGY

BAILUS WALKER, JR. *(Chair)*, Howard University Medical Center and American Public Health Association, Washington, D.C.
MELVIN E. ANDERSEN, Colorado State University, Fort Collins
EDWARD C. BISHOP, Parsons Corporation, Fairfax, Virginia
GARY P. CARLSON, Purdue University, West Lafayette, Indiana
JANICE E. CHAMBERS, Mississippi State University, Mississippi State
LEONARD CHIAZZE, JR., Georgetown University, Washington, D.C.
JUDITH A. GRAHAM, American Chemistry Council, Arlington, Virginia
SIDNEY GREEN, Howard University, Washington, D.C.
MERYL KAROL, University of Pittsburgh, Pittsburgh, Pennsylvania
STEPHEN U. LESTER, Center for Health Environment and Justice, Falls Church, Virginia
DAVID H. MOORE, Battelle Memorial Institute, Bel Air, Maryland
CALVIN C. WILLHITE, State of California, Berkeley

Staff

KULBIR S. BAKSHI, Program Director
SUSAN N. J. MARTEL, Senior Program Officer
ABIGAIL E. MITCHELL, Program Officer
KELLY CLARK, Assistant Editor
AIDA NEEL, Senior Project Assistant
TAMARA DAWSON, Project Assistant

BOARD ON ENVIRONMENTAL STUDIES AND TOXICOLOGY[1]

---

[1]This study was planned, overseen, and supported by the Board on Environmental Studies and Toxicology.

OTHER REPORTS OF THE BOARD ON ENVIRONMENTAL STUDIES AND TOXICOLOGY

Estimating the Public Health Benefits of Proposed Air Pollution Regulations (2002)
Biosolids Applied to Land: Advancing Standards and Practices (2002)
Ecological Dynamics on Yellowstone's Northern Range (2002)
The Airliner Cabin Environment and Health of Passengers and Crew (2002)
Arsenic in Drinking Water: 2001 Update (2001)
Evaluating Vehicle Emissions Inspection and Maintenance Programs (2001)
Compensating for Wetland Losses Under the Clean Water Act (2001)
A Risk-Management Strategy for PCB-Contaminated Sediments (2001)
Acute Exposure Guideline Levels for Selected Airborne Chemicals (2 reports; 2000, 2002)
Toxicological Effects of Methylmercury (2000)
Strengthening Science at the U.S. Environmental Protection Agency (2000)
Scientific Frontiers in Developmental Toxicology and Risk Assessment (2000)
Ecological Indicators for the Nation (2000)
Modeling Mobile-Source Emissions (2000)
Waste Incineration and Public Health (1999)
Hormonally Active Agents in the Environment (1999)
Research Priorities for Airborne Particulate Matter (3 reports, 1998-2001)
Ozone-Forming Potential of Reformulated Gasoline (1999)
Arsenic in Drinking Water (1999)
The National Research Council's Committee on Toxicology: The First 50 Years (1997)
Carcinogens and Anticarcinogens in the Human Diet (1996)
Upstream: Salmon and Society in the Pacific Northwest (1996)
Science and the Endangered Species Act (1995)
Wetlands: Characteristics and Boundaries (1995)
Biologic Markers (5 reports, 1989-1995)
Review of EPA's Environmental Monitoring and Assessment Program (3 reports, 1994-1995)
Science and Judgment in Risk Assessment (1994)
Pesticides in the Diets of Infants and Children (1993)
Dolphins and the Tuna Industry (1992)
Science and the National Parks (1992)
Human Exposure Assessment for Airborne Pollutants (1991)
Rethinking the Ozone Problem in Urban and Regional Air Pollution (1991)
Decline of the Sea Turtles (1990)

Copies of these reports may be ordered from The National Academies Press
(800) 624-6242 or (202) 334-3313
www.nap.edu

Other Reports of the Committee on Toxicology

Review of Submarine Escape Action Levels for Selected Chemicals (2002)
Standing Operating Procedures for Developing Acute Exposure Guideline Levels for Hazardous Chemicals (2001)
Evaluating Chemical and Other Agent Exposures for Reproductive and Developmental Toxicity (2001)
Acute Exposure Guideline Levels for Selected Airborne Contaminants, Volume 1 (2000), Volume 2 (2002)
Review of the US Navy's Human Health Risk Assessment of the Naval Air Facility at Atsugi, Japan (2000)
Methods for Developing Spacecraft Water Exposure Guidelines (2000)
Review of the U.S. Navy Environmental Health Center's Health-Hazard Assessment Process (2000)
Review of the U.S. Navy's Exposure Standard for Manufactured Vitreous Fibers (2000)
Re-Evaluation of Drinking-Water Guidelines for Diisopropyl Methylphosphonate (2000)
Submarine Exposure Guidance Levels for Selected Hydrofluorocarbons: HFC-236fa, HFC-23, and HFC-404a (2000)
Review of the U.S. Army's Health Risk Assessments for Oral Exposure to Six Chemical-Warfare Agents (1999)
Toxicity of Military Smokes and Obscurants, Volume 1(1997), Volume 2 (1999), Volume 3 (1999)
Assessment of Exposure-Response Functions for Rocket-Emission Toxicants (1998)
Toxicity of Alternatives to Chlorofluorocarbons: HFC-134a and HCFC-123 (1996)
Permissible Exposure Levels for Selected Military Fuel Vapors (1996)
Spacecraft Maximum Allowable Concentrations for Selected Airborne Contaminants, Volume 1
(1994), Volume 2 (1996), Volume 3 (1996), Volume 4 (2000)

Preface

In the 1980s, the U.S. Department of Defense (DOD) selected jet-propulsion fuel 8 (JP-8) as its primary fuel. JP-8 is widely used by the military not only for aircraft, but also for ground vehicles and other equipment, such as generators, cooking stoves, and tent heaters. Military personnel can be exposed to JP-8 vapors and aerosols during a number of operational scenarios, including aircraft refueling and maintenance. To protect the health of its personnel, DOD recommended an interim permissible exposure level (PEL) of 350 mg/m^3.

The Air Force requested that the National Research Council (NRC) review the available toxicologic, epidemiologic, and other relevant data on JP-8 and evaluate independently the scientific basis of the DOD's interim PEL of 350 mg/m^3 for JP-8. The NRC assigned this project to the Committee on Toxicology (COT), which assembled the Subcommittee on Jet-Propulsion Fuel 8 to prepare this report.

We thank the following Air Force personnel for providing valuable background information to the subcommittee: Brian Blazicko, Roger Gibson, John Hinz, David Mattie, James McDougal, and Richard Stotts. We also wish to express our gratitude to Geraldine Grant (George Mason University, Fairfax, Virginia), David Harris (University of Arizona, Tucson, Arizona), Glenn Ritchie (Geo-Centers, Inc., Wright-Patterson Air Force Base, Ohio), Mark Smulson (Georgetown University, Washington, D.C.), Steve Ullrich (M.D. Anderson Cancer Center, Houston, Texas), Russell White (Chevron Research and Technology Company, Richmond, California), and Mark Witten (University of Arizona, Tucson, Arizona) for providing background information to the subcommittee. We also thank Stephen Channel (U.S. Air Force), Thomas Neal (U.S. Air Force), and Kenneth Still (U.S. Navy) for their support of this project.

This report has been reviewed in draft form by persons chosen for their diverse perspectives and technical expertise, in accordance with procedures approved by the NRC's Report Review Committee. The purpose of this independent review is to provide candid, critical comments that will assist the institution in making its published report as sound as possible and to ensure that the report meets institutional standards for objectivity, evidence, and responsiveness to the study charge. The review comments and draft manuscript remain confidential to protect the integrity of the deliberative process. We wish to thank the following for their review of this report: Edward Bishop, Parsons Engineering Sciences, Inc., Fairfax, Virginia; Judith Graham, American Chemistry Council, Arlington, Virginia; Karl Kelsey, Harvard School of Public Health, Boston, Massachusetts; Carole Kimmel, U.S. Environmental Protection Agency, Washington, D.C.; Kannan Krishnan, University of Montreal, Quebec, Canada; David Lawrence, New York State Department of Health, Albany, New York; Judith MacGregor, Toxicology Consulting Services, Arnold, Maryland; Ceinwen Schreiner, C & C Consulting in Toxicology, Meadowbrook, Pennsylvania; and Bailus Walker, Jr., Howard University Medical Center, Washington, D.C.

Although the reviewers provided many constructive comments and suggestions, they were not asked to endorse the report's conclusions or recommendations, nor did they see the final draft of the report before its release. The review of this report was overseen by Dean Carter, University of Arizona, Tucson, who was appointed by the NRC to ensure that an independent examination of this report was conducted in accordance with institutional procedures and that all review comments were carefully considered. Responsibility for the final content of the report rests entirely with the subcommittee and the institution.

We are also grateful for the assistance of members of the NRC staff in the preparation of this report. The subcommittee acknowledges Abigail Mitchell, project director, and Kulbir Bakshi, program director of the Committee on Toxicology. Other staff members contributing to this report were James Reisa, director of the Board on Environmental Studies and Toxicology; Jessica Brock, senior project assistant; Norman Grossblatt, editor; and Kelly Clark, assistant editor.

Finally, we thank all members of the subcommittee for their expertise and dedicated effort throughout the study.

Melvin E. Andersen, PhD
Chair, Subcommittee on
Jet-Propulsion Fuel 8

Bailus Walker, Jr., PhD, MPH
Chair, Committee on Toxicology

Contents

Abbreviations

ACGIH	American Conference of Governmental Industrial Hygienists
ATSDR	Agency for Toxic Substances and Disease Registry
COT	Committee on Toxicology
CNS	central nervous system
DFM	diesel fuel marine
DNA	deoxyribonucleic acid
DOD	U.S. Department of Defense
FOB	functional observation battery
HDS	hydrodesulfurized
IARC	International Agency for Research on Cancer
JP-8	jet-propulsion fuel 8
LDH	lactate dehydrogenase
MDF	middle distillate fraction
MMAD	mass mean aerodynamic diameter
MN	micronucleus
NATO	North Atlantic Treaty Organization
NIOSH	National Institute for Occupational Safety and Health
NOAEL	no-observed-adverse-effect level
NRC	National Research Council
NTP	National Toxicology Program

OR	odds ratio
PEL	permissible exposure level
PB-PK model	physiologically based pharmacokinetic model
RD_{50}	respiratory depression in 50% of the animals tested
REL	recommended exposure limit
SCE	sister chromatid exchange
TOMM	test of memory and motivation
TWA	time-weighted average
UDS	unscheduled DNA synthesis
USAF	U.S. Air Force

Toxicologic Assessment of Jet-Propulsion Fuel 8

Summary

The U.S. Department of Defense (DOD) has, for many years, faced the logistical problem of using a variety of fuels for its aircraft, ground vehicles, and other equipment, such as cooking stoves and tent heaters. In the 1980s, DOD decided to initiate a 20-year (yr) fuel conversion process, in which most fuel-requiring equipment would be converted to exclusive use of jet propulsion fuel-8 (JP-8). With DOD and North Atlantic Treaty Organization (NATO) forces using an estimated 5 billion gallons of JP-8 each year, there is widespread exposure of DOD and NATO military personnel to JP-8.

In 1996, a previous subcommittee of the National Research Council's Committee on Toxicology (COT)[1] judged that the Navy's interim 8-hr time-weighted-average permissible exposure level (PEL) of 350 mg/m^3 for JP-4, JP-5, and JP-8 was adequate to protect the health of Navy personnel occupationally exposed to vapors from those fuels, based on the data available at that time; however, it identified a number of data gaps and recommended that the PEL for the three jet-fuel vapors be considered interim until further research had been completed.

Since the release of the 1996 report, considerable data on JP-8 have been generated. Because JP-8 is now being used more widely by DOD, and be-

[1]National Research Council. 1996. *Permissible Exposure Levels for Selected Miltiary Fuel Vapors*. Washington, DC: National Academy Press.

cause occupational exposures to JP-8 vapors and aerosols are known to occur, the Air Force requested that the National Research Council (NRC) again review the available toxicologic, epidemiologic, exposure, and other relevant data on JP-8; independently reevaluate the scientific basis of the PEL of 350 mg/m^3 for JP-8; identify data gaps; and make recommendations for future research relevant to deriving the PEL. The NRC assigned this project to the COT Subcommittee on Jet-Propulsion Fuel 8, which prepared this report.

THE SUBCOMMITTEE'S APPROACH TO ITS CHARGE

To address its charge, the subcommittee reviewed data on physical and chemical properties of JP-8, toxicokinetics of JP-8, epidemiologic and toxicologic evidence of adverse health effects of JP-8, and Air Force operational scenarios that might result in exposure to JP-8 vapors and aerosols. In addition to reviewing health-effects data on JP-8, the subcommittee reviewed toxicity data on kerosene and other kerosene-based fuels (such as JP-5) that are similar to JP-8. The subcommittee reviewed toxicity data on JP-8 vapors as well as JP-8 aerosols. The subcommittee used the available data to evaluate the scientific basis of the Air Force interim PEL of 350 mg/m^3.

CONCLUSIONS

The health-effects data on JP-8 and related fuels were reviewed for the following end points: respiratory tract toxicity, neurotoxicity, immunotoxicity, liver toxicity, kidney toxicity, reproductive and developmental toxicity, cardiovascular toxicity, genotoxicity, and carcinogenicity. JP-8 was found to be potentially toxic to the immune system, respiratory tract, and nervous system at exposure concentrations near the interim PEL of 350 mg/m^3. Those toxicities are summarized below.

Immune System

No immunotoxic effects were found in a study in which F344 rats and C57BL/6 mice were exposed to JP-8 *vapors* at concentrations up to 1,000 mg/m^3 on a continuous basis for 90 days. However, in other studies, inhalation exposure of C57BL/6 mice to JP-8 *aerosols* at a concentration of 100 mg/m^3 for 1 hr/day for 7 days led to decreased cellularity of the thymus;

exposure at 500 mg/m^3 for 1 hr/day for 7 days led to decreased spleen weight and cellularity; and exposure at 1,000 mg/m^3 for 1 hr/day for 7 days led to decreased ability of spleen cells to mediate several immune responses. The subcommittee reviewed the methods used to generate the exposure atmospheres in the studies using JP-8 aerosols and suspects that the total JP-8 concentrations in the atmosphere may have been underreported. However, even if the actual concentration was 10 times as high as the lowest concentration at which effects were observed (100 mg/m^3) (that is, if exposure was at a concentration of 1,000 mg/m^3), the observation of positive effects from a short exposure duration (1 hr/day for 7 days) at that concentration leads the subcommittee to conclude that the interim PEL of 350 mg/m^3 might be too high to be protective of human health (assuming the application of commonly used uncertainty factors). Because the positive results from JP-8 exposure in immunotoxicity assays were reported from only one laboratory, the subcommittee strongly recommends further research be conducted to validate those findings (see section on research recommendations). Results from those studies would provide data for establishing the PEL for JP-8 with greater confidence.

Respiratory Tract

No respiratory tract effects were found in F344 rats and C57BL/6 mice exposed to JP-8 vapors at concentrations of 500 or 1,000 mg/m^3 for 90 days. However, several animal studies conducted in F344 rats and C57BL/6 mice suggest that mixtures of JP-8 vapors and aerosols can result in pulmonary inflammation and alterations in pulmonary functions; such effects have been reported in C57BL/6 mice exposed at concentrations as low as 50 mg/m^3 for 1 hr per day for 7 days. As in the immune-system studies described above, the subcommittee suspects that the JP-8 concentrations in these studies may have been underreported. However, even if the actual concentration was 20 times as high (that is, if exposure was at a concentration of 1,000 mg/m^3), the observation of positive effects from a short exposure duration (1 hr/day for 7 days) at that concentration leads the subcommittee to conclude that the interim PEL of 350 mg/m^3 might be too high to be protective of human health (assuming the application of commonly used uncertainty factors). Because the positive results from JP-8 exposure in respiratory toxicity assays were reported from only one laboratory, the subcommittee strongly recommends further research be conducted to validate those findings (see section on research recommendations). Results from those studies would provide data for establishing the PEL for JP-8 with greater confidence.

Nervous System

Animal studies have investigated the effects of several jet fuels on a number of neurobehavioral end points. Neurobehavioral effects were reported in Sprague-Dawley and F344 rats exposed to JP-8 or JP-5 vapors at concentrations of 1,000 mg/m^3 for 6 hr/day, 5 days/wk for 6 wk or to JP-8 aerosols at concentrations of 1,059 mg/m^3 for 1 hr/day, 5 days/wk for 4 wk. The relevance to humans of the toxicity end points evaluated in those studies is not known and no dose-response relationships were demonstrated in the studies. Furthermore, those positive findings need to be validated against other well-established neurotoxicity end points. However, the findings provide further indication that the interim PEL of 350 mg/m^3 might be too high to be protective of human health.

Cancer

The carcinogenicity of JP-8 has not been investigated in epidemiologic studies or in chronic lifetime inhalation-exposure studies in experimental animals. Ninety-day continuous inhalation-exposure studies of JP-5 have been conducted in C57BL/6 mice exposed at a concentration of 750 mg/m^3, and no increase in the incidence of tumors was observed. The carcinogenicity data available on mixtures similar to JP-8 (such as other jet fuels and middle distillates) indicate that most of these materials induce skin tumors in mice when topically applied in excessive amounts and under conditions of excessive skin irritation.

The subcommittee is aware of a suspected cluster of acute lymphocytic leukemias (ALL) in Fallon, Nevada, and that exposure to JP-8, originating from a naval base located in that town, is under investigation as a possible cause of the ALL cluster (exposures to other chemicals are being investigated as well). However, no scientific studies were found in the published literature that examined a potential relationship between ALL and JP-8 exposure; therefore, the subcommittee could not reach any conclusion concerning exposure to JP-8 and this suspected cancer cluster.

Other Toxicity End Points

The subcommittee also reviewed toxicologic and epidemiologic data on other end points: hepatotoxicity, renal toxicity, reproductive and developmental toxicity, cardiovascular toxicity, and genotoxicity of JP-8. No relevant adverse effects were observed for hepatotoxicity, renal toxicity, and cardiovascular toxicity, although the exposure concentration did not exceed 1,000 mg/m^3. Adequate studies have not been conducted to assess the toxicity potential of inhaled JP-8 for reproductive toxicity, developmental toxicity, and genotoxicity.

Subcommittee's Evaluation of the Interim PEL of 350 mg/m^3 for JP-8

On the basis of the available toxicologic data, the subcommittee concludes that the interim PEL of 350 mg/m^3 for JP-8 proposed by the Air Force might be too high to be protective of human health. It is beyond the charge to the subcommittee to propose a specific PEL for JP-8; such decisions necessarily involve more than scientific considerations. In addition, further studies on JP-8 are necessary to provide the requisite data to establish a PEL with greater confidence. Because JP-8 vapors and aerosols have different toxic potencies, the Air Force should consider developing separate PELs for vapors and aerosols.

The subcommittee further concludes that in addition to inhalation exposures, the potential exists for a substantial contribution to the overall JP-8 exposure by the dermal route, including mucous membranes and the eyes, either by contact with vapors and aerosols or by direct skin contact with JP-8. It should be noted that earlier this year, the American Conference of Governmental Industrial Hygienists proposed a Threshold Limit Value for kerosene and jet fuels, as a total hydrocarbon vapor, of 200 mg/m^3.[2] Also, ExxonMobil Biomedical Sciences, Inc., has set an occupational exposure level of 5 mg/m^3 for kerosene and middle distillate fuel aerosols.[3]

RESEARCH RECOMMENDATIONS

Because findings from several studies indicate the potential for adverse health effects from exposure to JP-8 aerosols at concentrations below the interim PEL of 350 mg/m^3 and because the JP-8 vapor concentrations tested were approximately 1,000 mg/m^3 (that is, less than three times the interim PEL), the subcommittee strongly recommends that a battery of inhalation-toxicity tests in experimental animals be conducted with JP-8 vapors and mixtures of vapors and aerosols. The animal studies should include evalua-

[2]ACGIH (American Conference of Governmental Industrial Hygienists). 2002. Threshold Limit Values and Biological Exposure Indices. Cincinnati, Ohio.

[3]ExxonMobil Biomedical Sciences, Inc. 2001. ExxonMobil Occupational Exposure Limits for Chemical Contaminants. Annandale, New Jersey.

tions of immune, nervous system, respiratory, hepatic, renal, cardiovascular, reproductive, developmental, and in vivo genetic toxicity end points, including basic evaluations of clinical effects and histopathology.

Because the composition of JP-8 varies from batch to batch, scientists with expertise in petroleum toxicology should be consulted to design the best approach for testing the toxicity of JP-8 in animal studies (for example, testing JP-8 samples at the extremes of their composition ranges or testing JP-8 samples so that the concentrations of component classes can be correlated with toxic end points).

Animal studies involving exposures to aerosols should be designed in collaboration with scientists knowledgeable in aerosol generation, aerosol physics, and atmospheric quantification of vapors and aerosols to ensure accurate characterization of exposure atmospheres. The exposure conditions in the animal studies should mimic exposures encountered in the workplace (such as proper ratios of vapors to aerosols).

Because inhalation exposures greater than approximately 1,000 mg/m^3 for pure JP-8 vapors are difficult to achieve, the Air Force should consider conducting studies using saturated vapor atmospheres on larger numbers of animals or employ longer exposure durations (that is, longer than 90 days) to increase the power of the studies for observing adverse effects in various organ systems.

The lifetime carcinogenicity of JP-8 has not been studied; therefore, the subcommittee recommends that 2-yr inhalation carcinogenicity bioassays be conducted in two experimental animal species.

The subcommittee recommends that human blood samples from JP-8-exposed persons be assayed for indicators of immunotoxicity to determine whether effects observed in experimental animals are also observed in humans.

Preliminary positive findings were reported in two neurologic tests (eye-blink and postural-sway tests) conducted as part of a recent Air Force human study. The subcommittee recommends that results from those two tests be validated with standard neurologic tests.

The subcommittee also recommends that toxicokinetic studies be conducted so that existing human studies on JP-8 and related fuels can be better interpreted. Those studies should provide quantitative information on the relationship of blood and tissue concentrations of JP-8 components after vapor and aerosol exposures to JP-8. Traditional compartmental and physiologically-based toxicokinetic models that take into account absorption, distribution, metabolism, and elimination should include studies on JP-8 and on longer-chain *n*-alkanes, naphthalene, benzene, and other components of JP-8. With improved dosimetry, available human data from a recently completed Air Force human study should be reevaluated on the basis of body burden of JP-8.

The Air Force should conduct studies to estimate exposures of its personnel to JP-8 vapors or mixtures of vapors and aerosols. Health-effect assessments and blood analysis for JP-8 components should be conducted in conjunction with exposure assessments so that correlations between actual exposures and adverse effects can be made. Those data are likely to be useful for validating any toxicokinetic modeling based on rodent toxicity studies.

The subcommittee recommends that dermal exposures to Air Force personnel in some occupational settings (such as maintenance of aircraft fuel tanks) be minimized by the use of appropriate protective clothing or other measures. It also recommends that DOD evaluate the effectiveness of various protective clothing for personnel who are likely to come into contact with JP-8 and that it use the most effective protective clothing. Furthermore, the subcommittee recommends that other industrial hygiene practices be instituted to reduce or prevent exposures to JP-8 vapors or aerosols.

1

Introduction

Jet-propulsion fuel 8 (JP-8) is a kerosene-based multipurpose fuel that is in wide use by the U.S. military. The military is in the process of converting to JP-8 for use in all its aircraft (except Navy ship-based aircraft, which will continue to use JP-5), ground vehicles, and support equipment, such as generators, cooking stoves, and tent heaters (Makris 1994; Edwards et al. 2001).

There are several reasons for the conversion: using a single fuel will eliminate many logistical problems associated with transporting and distributing multiple fuels to various bases within the United States and to U.S. operations in other countries (Makris 1994); JP-8 is produced from jet fuel A and jet fuel A-1, which are used in commercial aircraft and are readily available throughout the world (Makris 1994; USAF 1996; Chevron 2000); it has a higher flash point than several other jet fuels used by the military (such as wide-cut jet fuels, which are mixtures of gasoline and kerosene) and is therefore less flammable and less likely to ignite accidentally (Zeiger and Smith 1998); and it has a lower vapor pressure than wide-cut jet fuels, so less fuel is lost to evaporation (Makris 1994). The U.S. Department of Defense (DOD) identified JP-8 as its single military fuel in the 1980s, but the conversion has been gradual because of the need to modify engines and other equipment (Zeiger and Smith 1998). The conversion to JP-8 is scheduled to take approximately 20 years.

The U.S. military services and the North Atlantic Treaty Organization forces use an estimated 5 billion gallons of JP-8 each year (Zeiger and Smith

1998; Henz 1998). Military personnel are exposed to JP-8 during aircraft fueling and maintenance operations. Because of JP-8's comparatively low volatility and higher flash point than some other jet fuels (e.g., JP-5), jet engines powered by it do not start as easily or burn fuel as completely, particularly under cold conditions, as they do when powered by wide-cut jet fuels. Cold-engine starts are known to produce plumes of unburned aerosolized fuel. Workers involved with fueling operations on aircraft are exposed to JP-8 vapor, aerosol, and liquid during startup procedures. There are also anecdotal reports from exposed workers of dizziness, skin irritation, and of smelling and tasting the fuel hours after exposure (Zeiger and Smith 1998). For additional general information on JP-8, consult the Agency for Toxic Substances and Disease Registry's *Toxicological Profile for Jet Fuels (JP-5 and JP-8)* (ATSDR 1998).

DOD recommended an interim permissible exposure level (PEL) for JP-8 of 350 mg/m^3 (NRC 1996). A PEL is an allowable time-weighted exposure concentration in workplace air averaged over an 8-hr shift. No other national agencies or organizations have recommended regulations or guidelines applicable to JP-8. Two agencies have established regulations for petroleum distillates: the National Institute for Occupational Safety and Health (NIOSH) set an 8-hr recommended exposure limit (REL) time-weighted average (TWA) of 350 mg/m^3, and the Occupational Safety and Health Administration (OSHA) set a PEL TWA of 2,000 mg/m^3 (NIOSH 1997; OSHA 29 CFR 1910.1000 [1997]). NIOSH has established a REL TWA of 100 mg/m^3 for kerosene (NIOSH 1997). The American Conference of Governmental Industrial Hygienists (ACGIH) recently proposed a Threshold Limit Value for kerosene and jet fuels (as a total hydrocarbon vapor) of 200 mg/m^3 (ACGIH 2002). ExxonMobil Biomedical Sciences, Inc., has set occupational exposure levels for kerosene and other middle distillate fuels of 500 mg/m^3 for vapors and 5 mg/m^3 for aerosols (ExxonMobil Biomedical Sciences, Inc. 2001). The International Agency for Research on Cancer concluded that jet fuel is "not classifiable" as to its carcinogenicity in humans (IARC 1989). ACGIH classified kerosene and jet fuels as "confirmed animal carcinogens with unknown relevance to human skin" (ACGIH 2002).

SUMMARY OF 1996 NATIONAL RESEARCH COUNCIL REPORT ON MILITARY FUELS

In 1996, the National Research Council (NRC) released the report of the Committee on Toxicology (COT) Subcommittee on Permissible Exposure Levels for Military Fuels, which evaluated DOD's interim PEL of 350 mg/m^3 by reviewing data on the toxicity of the vapors from JP-4, JP-5, JP-8, and diesel fuel marine in experimental animals and humans (NRC 1996). The executive summary from that report is presented as Appendix A.

The Subcommittee on Permissible Exposure Levels for Military Fuels judged that, on the basis of available data, DOD's PEL of 350 mg/m^3 for the fuel vapors is adequate to protect the health of naval personnel exposed to them occupationally (NRC 1996). However, because of uncertainties in the database, the PEL should still be considered interim until further research has been completed. The subcommittee recommended that data be obtained on exposures during operational procedures, including exposure to respirable aerosols of unburned fuels; that studies be conducted on the possible effects of high-level acute and low-level chronic exposure to fuel vapors on the central nervous system; and that research be conducted on the effect of fuel vapors on hepatotoxicity in experimental animals to help to identify a no-observed-adverse-effect level for JP-8 with greater confidence.

THE CHARGE TO THE SUBCOMMITTEE

Since the release of the 1996 NRC report, additional data on JP-8 have been generated. In light of those data, the U.S. Air Force asked the NRC to review the toxicologic, epidemiologic, and other relevant data on JP-8 vapors and aerosols to assess the scientific basis of the interim PEL of 350 mg/m^3 proposed by DOD, identify data gaps, and make recommendations for future research relevant to deriving the PEL. The NRC assigned the project to COT and assembled the Subcommittee on Jet-Propulsion Fuel 8, which prepared the present report.

THE SUBCOMMITTEE'S APPROACH TO ITS CHARGE

The subcommittee reviewed information regarding the physical and chemical properties of JP-8, military operational scenarios that might result in exposures to fuel vapors and aerosols, toxicokinetics of the fuel, and epidemiologic and toxicologic evidence of adverse health effects of exposures to JP-8 vapors and aerosols. Because JP-8 is a kerosene-based fuel and its toxicologic properties are thought to be similar to those of kerosene, the subcommittee also reviewed toxicity data on kerosene and other kerosene-based fuels. The subcommittee used the body of information on JP-8, kerosene, and other kerosene-based fuels to evaluate the interim PEL of 350 mg/m^3 and determine whether it is adequate to protect the health of military personnel exposed to JP-8 occupationally.

PHYSICAL AND CHEMICAL PROPERTIES OF JP-8

JP-8 is a complex mixture containing more than 200 aliphatic and aromatic hydrocarbon compounds with nine to 17 (or perhaps more) carbon atoms, including thousands of isomeric forms that distill at 170-325°C, and three to six nonhydrocarbon performance additives (Henz 1998; DOD 1992). The precise composition of JP-8 varies from batch to batch. Some of the physical and chemical properties of JP-8 are summarized in Table 1-1, and the additives in JP-8 are summarized in Table 1-2.

The hydrocarbon portion of jet fuels is made from low-sulfur or desulfurized distillate kerosene streams, usually blended with cracked or hydrocracked heavier streams to produce a fuel that meets specific performance specifications. JP-8 is an extremely complex mixture with specifications established for boiling point, sulfur (maximal percentage, 0.3%), olefins (maximal percentage, 5.0%), and aromatics (maximal percentage, 22%) (Vere 1984). The aromatics limit is primarily to prevent excessive smoke production during combustion. Aside from the aromatics, most of the remainder of JP-8 consists of the *n*-alkanes, isoalkanes, and naphthenics component classes. There is no minimal requirement for the aromatics class, and aromatics are generally not desired in jet fuels. In addition to hydrocarbons, jet fuel contains small amounts of sulfur and nitrogen as heterocyclic substituents generally in structures containing one or two rings. At a boiling point of about 500°F, an atmospheric petroleum distillate stream (kerosene) has about 4,000 different *n*-alkanes and isoalkanes. Combinations of the naphthenes, aromatics, and heterocyclics are also present, and the total number of components is very large.

The approximate ranges of the major hydrocarbon classes (by volume %) in JP-8 are as follow (Vere 1984):

n-alkanes + isoalkanes	33-61%
olefins	0.5-5%
naphthenes (naphthenics)	10-45%
aromatics	12-22%

ORGANIZATION OF THIS REPORT

This report contains 11 chapters in addition to this introductory chapter. Chapter 2 describes issues relevant to assessing exposure of military personnel to JP-8. Chapter 3 discusses the toxicokinetics and toxicody-namics of JP-8. Chapters 4-10 summarize studies on the effects of JP-8 on the respiratory

TABLE 1-1 Physical and Chemical Properties of JP-8

Molecular weight:	» 180
Synonyms:	Jet fuel JP-8, kerosene, aviation kerosene, fuel oil no. 1, jet kerosene, turbo fuel A, straight-run kerosene, distillate fuel oil–light, MIL-T-83133B, AVTUR, NATO F-34
CAS registry number:	8008-20-6 (kerosene)/70892-10-3 (fuel oil 1)
Freezing point, maximum:	-47°C
Boiling point:	175-300°C
10% recovered, maximum:	205°C
End point, maximum:	300°C
Flash point, minimum:	38°C
Vapor pressure:	0.52 mm Hg (10°C), 1.8 mm Hg (28°C)
Specific gravity, kg/L, 15°C,	
minimum:	0.775
maximum:	0.840
Heating value, Btu/lb,	
minimum:	18,400
Viscosity, maximum at -20°C:	8
Physical state:	Liquid
Color:	Clear and bright
Solubility in water:	5 mg/L (kerosene)
Vapor density (air = 1):	4.5-5
Liquid density (water = 1):	0.788-0.845 kg/L
Odor:	Kerosene-like
Conversion factors at standard temperature and pressure:	1 ppm = 8.0 mg/m^3 1 mg/m^3 = 0.125 ppm

Sources: NRC 1996; ATSDR 1998.

tract, the nervous system, the immune system, the liver, the kidney, reproduction and development, and the cardiovascular system. Chapters 11 and 12 provide information on the genotoxic and carcinogenic effects of exposure to JP-8, respectively. This report also contains three appendixes. Appendix A

TABLE 1-2 Additives in Jet-Propulsion Fuel 8

Additive[a]	Function	Quantity	Required or Optional
Diethylene glycol monomethyl ether (DiEGME)	Ice inhibition	0.1 vol/vol %	Required
Stadis 450	Static inhibition	2 mg/L	Required
DCI-4A	Corrosion inhibition	15 mg/L	Required
Antioxidant	Inhibition of gum formation	25 ppm	Optional
Metal deactivator	Control of metal-catalyzed fuel deterioration	3 ppm	Optional

[a]Stadis 450 and DCI-4A are proprietary formulations; the antioxidant is *N,N*-diisopropylparaphenylene diamine or various blends of hindered phenols; the metal deactivator is *N,N*-disalicylidene-1,2-propanediamine or *N,N*-disalicylidene-1,2-cyclohexanediamine.
Source: Major Tom Miller, U.S. Air Force, Wright-Patterson Air Force Base, Ohio.

contains the executive summary from the 1996 Research Council report *Permissible Exposure Levels for Selected Military Fuel Vapors;* Appendix B contains the executive summary and introduction of *JP-8 Final Risk Assessment*, an unpublished report summarizing a recent Air Force-funded human-health study on JP-8; and Appendix C reviews types of tests used to assess neurologic function in humans after exposure to JP-8.

REFERENCES

ACGIH (American Conference of Governmental Industrial Hygienists). 2002. Threshold Limit Values and Biological Exposure Indices. American Conference of Governmental Industrial Hygienists, Cincinnati, OH.

ATSDR (Agency for Toxic Substances and Disease Registry). 1998. Toxicological Profile for Jet Fuels (JP-5 and JP-8). U.S. Department of Health and Human Services, Public Health Service, Agency for Toxic Substances and Disease Registry, Atlanta, GA.

Chevron. 2000. Technical Review of Aviation Fuels. San Ramon, CA: Chevron Products Company.

DOD (U.S. Department of Defense). 1992. Military Specifications: Turbine Fuel, Aviation, Grades JP-4, JP-5, and JP-5/JP-8 ST. MIL-T5624P.

Edwards, R., B. Harrison, and L. Maurice. 2001. Properties and Usage of Air Force Fuel: JP-8. AIAA 2001-0498. Presentation at the 39th Aerospace Meeting and Exhibit, Reno, NV, Jan. 8-11, 2001. American Institute of Aeronautics and Astronautics, Inc., Reston, VA.

ExxonMobil Biomedical Sciences, Inc. 2001. ExxonMobil Occupational Exposure Limits for Chemical Contaminants. ExxonMobil Biomedical Sciences, Inc., Annandale, New Jersey.

Henz, K. 1998. Survey of Jet Fuels Procured by the Defense Energy Support Center, 1990-1996. Defense Logistics Agencies, Ft. Belvior, VA.

IARC (International Agency for Research on Cancer). 1989. Occupational Exposures in Petroleum Refining, Crude Oil and Major Petroleum Fuels. IARC Monographs on the Evaluation of Carcinogenic Risks to Humans, Vol. 45. Lyon: International Agency for Research on Cancer, World Health Organization.

Makris, N.J. 1994. JP-8: A conversion update. Flying Safety 50(10):12-13.

NIOSH (National Institute for Occupational Safety and Health). 1997. NIOSH Pocket Guide to Chemical Hazards. DHHS (NIOSH) 97-140. U.S. Dept. of Health and Human Services, Public Health Service, Centers for Disease Control and Prevention, National Institute for Occupational Safety and Health, Cincinnati, OH.

NRC (National Research Council). 1996. Permissible Exposure Levels for Selected Military Fuel Vapors. Washington, DC: National Academy Press.

USAF (U.S. Air Force). 1996. History of Aviation Fuel Development in the U.S. Air Force Research Laboratory, Propulsion Directorate, Fuels Branch, U. S. Air Force.

Vere, R.A. 1984. Aviation fuels. Pp. 723-771 in Modern Petroleum Technology, Part 2, 5th Ed., G.D. Hobson, ed. Chichester: John Wiley & Sons.

Zeiger, E., and L. Smith. 1998. The first international conference on the environmental health and safety of jet fuel. Environ. Health Perspect. 106(11):763-764.

2

Dosimetry and Exposure Assessment of Jet-Propulsion Fuel 8

This chapter discusses issues relevant to assessing exposure of military personnel to jet-propulsion fuel 8 (JP-8). The chapter begins with a description of various scenarios under which military personnel are exposed to JP-8, followed by a brief discussion of the challenges of quantifying human exposure to this distillate fuel. The next section contains a summary of data from studies that have measured concentrations of several components of JP-8 in ambient air at Air Force aircraft maintenance sites. Studies measuring body burden of several JP-8 components in workers involved in aircraft maintenance are also presented. The final section of this chapter describes how the physical and chemical properties of JP-8 affect uptake into the body from exposure by the inhalation, dermal, and oral routes. This last section also serves as a prelude to interpretation of animal toxicity studies conducted with distillate fuels (e.g., JP-8) that are described in later chapters.

BACKGROUND

Henz (1998) recently estimated that the U.S. Department of Defense (DOD) and North Atlantic Treaty Organization (NATO) partners use 5 billion gallons of JP-8 per year for powering aircraft and land-based military vehicles. Quantitatively and qualitatively, JP-8 is probably responsible for the most common and abundant potential chemical exposure of DOD and NATO personnel. Occupational exposure of military and civilian personnel to JP-8 may occur in the following settings (see Table 2-1): transportation and storage of JP-8 fuel; aircraft fueling and defueling; maintenance of aircrafts; cold engine starts and performance testing; and operation and maintenance of other Air Force equipment and machinery. Air Force personnel working in aircraft fuel-cell maintenance shops and fuels-specialty and fuels-transportation shops are probably at the greatest risk for exposure to JP-8 (ATSDR 1998; R. Gibson, U.S. Air Force, personal communication, 2001). Additional exposure scenarios include application of JP-8 in fueling military tent heaters, use of fuel as an aircraft heat sink, and cleaning and degreasing with fuel (CDC 1999; Zhou and Cheng 2000; Cheng et al. 2001). During the Persian Gulf War, JP-8 was routinely used to control and suppress desert sand, and combusted JP-8 fuel was used to obscure troops and equipment (CDC 1999). With desert surface temperatures commonly exceeding 120°F, substantial exposure may have occurred as a result of vaporization of JP-8. When vaporized jet fuel mixes with wind-blown ultrafine desert sand particles, pulmonary exposure is highly possible.

ASSESSMENT OF OCCUPATIONAL EXPOSURE TO JP-8

Deliberations on the scientific basis of the interim permissible exposure level (PEL) of 350 mg/m^3 for JP-8 required the subcommittee to review relevant exposure assessment, epidemiologic, and toxicologic data. The studies all involved exposure to JP-8 or similar compounds; however, in many studies it is a challenge to qualitatively and quantitatively assess the exposure and dose of components of JP-8. Measurement of occupational exposures in various settings is problematic because JP-8 is a complex mixture of chemicals and personnel may be exposed to vapors, aerosols, or both, depending on the workplace setting. Furthermore, there are no standardized industrial hygiene sampling methods or analytical methods for JP-8. Ambient concentrations have been measured with standard industrial-hygiene methods to quantify some components of JP-8 (mostly aromatic substances, including benzene, naphthalene, toluene, and xylene). Given that the major chemical constituents of JP-8 are C9-C17+ aliphatic hydrocarbons, it is unclear how these aromatic components represent a true measure of total occupational exposure to JP-8.

TABLE 2-1 Major JP-8 Exposure Scenarios

Exposure Scenario	DOD Personnel	Exposure Type
Cold starts	Ground crew	Aerosol, vapor, liquid
Fueling, defueling	Fuelers	Aerosol, vapor, liquid
Engine, fuel cells	Maintenance workers	Vapor, liquid
Repair	Maintenance workers	Vapor, liquid
Fuel transport	Fuel handlers	Aerosol, vapor, liquid
Stoves, heaters	General troops	Vapor, liquid
Spills	Cleanup crews	Vapor, liquid
Tanks, ground vehicles	General troops	Vapor, liquid

Source: Witten 2002. Reprinted with permission of the author.

JP-8 Concentrations in Ambient Air

Pleil et al. (2000) reported on the collection of area breathing-zone samples taken at fuel maintenance operation sites at a number of domestic Air Force bases with commercially available, compact, battery-operated, personal whole-air samplers. The ambient air concentrations were 2,987 parts per billion (ppb) for benzene, 16,026 ppb for toluene, 9,588 ppb for ethylbenzene, 20,993 ppb for xylenes, 34,138 ppb for nonane, 31,344 ppb for decane, and 31,007 ppb for undecane (Pleil et al. 2000). Carlton and Smith (2000) measured JP-8 (based on total hydrocarbons) and benzene vapor concentrations during aircraft fuel-tank entry and repair at 12 Air Force bases. The highest 8-hr time-weighed average (TWA) fuel exposure was 1,304 mg/m^3; the highest 15-min short-term exposure was 10,295 mg/m^3. Overall, workers who repaired fuel tanks that contained explosion-suppression foam had the highest exposure to JP-8 total hydrocarbons compared with workers who repaired fuel tanks without foam. The highest benzene exposure in workers involved in the repair of foamed fuel tanks was 49.1 μg/m^3. Fuel tanks with cross ventilation had much lower concentrations of JP-8 than fuel tanks with poor ventilation. In an Australian study, JP-8 vapor concentrations were found to reach 2,823 mg/m^3 inside wet B747 aircraft fuel tanks (Yeung et al. 1997).

Recently, the Air Force assessed the potential health effects of acute exposure to JP-8 (TIEHH 2001; see Appendix B). The subjects were selected from several Air Force bases and placed in two groups: workers who were routinely exposed to JP-8 as part of their Air Force Specialty Code (workers involved in maintenance of aircraft-fuel cells were considered highly exposed and workers employed in fuels-transportation shops were considered moderately exposed) and age- and sex-matched unexposed referents (Air Force active-duty personnel with no direct contact with JP-8). JP-8 exposed workers entering aircraft-fuel cells wore respirators and coveralls, which are permeable and absorbent, so there was an opportunity for substantial dermal exposure to JP-8. Preliminary results indicate that the median concentration of naphthalene in air measured in the breathing zone of subjects in the low-exposure category (workers with no routine contact with JP-8) was 1.9 $\mu g/m^3$, about 4 times the concentration in ambient air (Egeghy and Rappaport 2001). The median benzene concentration was 3.1 $\mu g/m^3$ (2-fold less than global outdoor benzene concentration). In the moderate-exposure group, the median naphthalene and benzene concentrations were 10.4 and 7.45 $\mu g/m^3$, respectively. In the high-exposure group, the concentrations of those chemicals were at least 30 times higher than for the moderate-exposure group.

Overall, JP-8 air concentrations according to surrogate markers, such as total hydrocarbons, naphthalene, and benzene, appeared to be highest in aircraft fuel tanks, especially those containing explosion-suppression foam (Carlton and Smith 2000; Egeghy and Rappaport 2001). The increased air concentrations appear to result from the tendency of foam to absorb fuel. At elevated temperatures inside the tank, the fuel volatilizes to yield higher air concentrations of JP-8. Fuel tanks with appropriate cross ventilation have much lower interior air concentrations of JP-8.

Estimates of Dermal Exposure

Skin can be an important route of JP-8 exposure. Aircraft fuel-maintenance workers may be exposed to liquid jet fuel for more than 10 min, which gives ample opportunity for dermal exposure. Except for chemical-resistant eyewear, footwear, gloves, and cotton coverall jumpsuits, which are permeable to JP-8, there is little protection of skin. Prolonged JP-8 skin contact can induce irritation, contact dermatitis, and sensitization (Wolfe et al. 1997; Ullrich 1999).

Nylander-French and Archer (2001) reported preliminary results of an acute dermal-exposure study conducted with Air Force fuel-cell maintenance workers. A noninvasive tape-stripping technique was used to measure JP-8 concentrations in skin using naphthalene as a surrogate marker of exposure to the whole fuel. The technique permits the estimation of the dermal retention of JP-8 after the removal of stratum corneum by successive tape stripping with an adhesive tape. The high-exposure group (workers involved in maintenance of aircraft-fuel cells) had the highest skin exposure but also had the highest variability (5 orders of magnitude). Compared with the high-exposure group, exposure in the moderate- and low-dose groups (workers employed in fuels-transportation shops and unexposed referents, respectively) followed a descending trend. The mean, median, and range of naphthalene exposure for all subjects were 35.7, 15.0, and less than the limit of detection to 25,287 ng per tape strip, respectively. That mean naphthalene mass did not change substantially between various strippings suggests that naphthalene was able to penetrate the stratum corneum. Such penetration can contribute to delayed and cumulative systemic absorption of some JP-8 components through skin. Given that over 200 hydrocarbons are present in JP-8 and that hydrocarbons differ in skin disposition, overall skin absorption of JP-8 components remains poorly characterized. The use of surrogate markers representing a few hydrocarbons is at most a qualitative assessment of dermal exposure to JP-8 fuels.

MEASUREMENT OF BODY BURDEN OF JP-8

Environmental monitoring of JP-8 is crude and can lead to incorrect information about body burdens of JP-8 constituents. In a recent preliminary Air Force acute-exposure study, JP-8 components were measured in blood in a highly exposed group (fuel-cell maintenance workers) and an unexposed referent group (Gibson et al. 2001). The surrogate exposure to JP-8 was assessed as a "JP-8 fingerprint" representing the combined concentration of various aliphatic hydrocarbons, including nonane, decane, undecane, and dodecane. The JP-8 fingerprint index represents 15% of the JP-8 vapor and 11% of the liquid vapor. In the high-exposure group of workers, the concentration range was extremely large with values ranging from 1 to 124 ng/mL (mean, 10.24 ng/mL; median, 6 ng/mL). In the referent group, the concentration in the JP-8 fingerprint was 0-45 ng/mL (mean, 2.15 ng/mL; median, near zero). In addition to interperson and intergroup variability, there was a large variability in exposure among workers at various Air Force bases. That could reflect differences in exposure and variation in fuel contents. Other factors affecting variability include the amount of lipid in blood, hemoglobin concentration, meal patterns, and body fat.

Pleil (2001) analyzed preexposure and postexposure breath samples for a number of single-ring hydrocarbons (such as benzene, toluene, ethylbenzene, *m*-xylene, 4-ethyltoluene, 1,3,5-trimethylbenzene, and styrene) and C9-C12 *n*-alkanes and identified them collectively as a JP-8 fingerprint. The JP-8 exposure measurement in expired air was considered an important indicator of aggregate exposure and may collectively represent the transient JP-8 body burden from inhalation and dermal exposure. As indicated earlier, there was a large variation among bases in exposure to JP-8. Preliminary results show that at one Air Force base in Little Rock, Arkansas, the sum of benzene, toluene, ethylbenzene, and xylenes (BTEX) measurements was about 14% of the JP-8 fingerprint value; at Pope Air Force Base in North Carolina, the BTEX sum was about 50% of the JP-8 fingerprint value. The ranges of postexposure values overlapped substantially between exposed and unexposed subjects. It is often difficult to distinguish between exposed and unexposed workers on the basis of the JP-8 fingerprint, and simple division into highly exposed, moderately exposed, and unexposed workers appears to be an inadequate way to correlate adverse health effects with JP-8 body burden. About 10% of subjects had a markedly elevated JP-8 body burden. The reasons for their high exposures were not known.

In addition to measurement of the JP-8 fingerprint, benzene and naphthalene were measured in end-exhaled breath and with passive personal monitors attached to the clothing just below the chin (Egeghy and Rappaport 2001). Preliminary results showed that median postexposure benzene concentrations in exhaled air were 4.6, 9.0, and 11.4 $\mu g/m^3$ for the low-, moderate-, and high-exposure groups, respectively, and overall range was less than 1.5 to 153 $\mu g/m^3$. Preliminary results showed that the median postexposure naphthalene concentrations in exhaled air were less than 0.73, 0.93, and 1.83 $\mu g/m^3$ in the low-, moderate-, and high-exposure groups, respectively, and the overall range was less than 0.5 to 15.8 $\mu g/m^3$. Overall, the median postexposure concentrations exceeded the median preexposure concentrations by about 2- to 3-fold. Naphthalene concentrations were higher than benzene concentrations in the environmental breathing-air samples; the opposite was true for exhaled-air samples. That could be related to lower volatility and higher expected blood:air partition coefficient for naphthalene. The best correlation between the environmental and exhaled-air concentrations was obtained for the high-exposure group (with large variability); however, this correlation was weak for the low- and moderate-exposure groups. Both correlations were stronger for naphthalene than for benzene.

Benzene, naphthalene, and hydroxynaphthalene metabolites were measured in urine before and after exposure (Serdar et al. 2001). Urinary concentrations of benzene and naphthalene may be good surrogates of internal exposure because they reflect inhalation and skin exposure. Preliminary results indicated that, in the high-exposure group, there was strong correlation between workplace air and exhaled-air concentrations of each of these chemicals. A high correlation was also observed for urinary naphthalene and its metabolite. Urinary concentrations of these chemicals were higher in smokers, but the effect of smoking became less significant as the exposure to JP-8 increased. That air and preexposure urine samples did not show a positive correlation suggests that preexposure urinary benzene and naphthalene concentrations were the result of sources other than occupational exposures.

PROTEIN ADDUCTS AS SURROGATE BIOMARKERS OF JP-8 EXPOSURE

Naphthalene in exhaled air and its metabolites in urine have been used as a surrogate marker of JP-8, although naphthalene constitutes only a small fraction of the hydrocarbon content of JP-8 (Serdar et al. 2001; Egeghy and Rappaport 2001). Because the majority of inhaled naphthalene is eliminated rapidly, this represents an assessment of only acute exposure to naphthalene and provides no information on long-term steady-state exposure to JP-8. Protein (specifically, albumin and hemoglobin) adducts of reactive metabolites of naphthalene (such as 1, 2-, and 1,4-naphthoquinones) were measured as possible surrogate markers of long-term JP-8 exposure in Air Force personnel involved in fuel-cell maintenance (Waidyanatha and Rappaport 2001). Preliminary results showed that 1,2-naphthoquinone albumin adduct (NQ-Ab) concentrations in blood were slightly greater in personnel exposed to JP-8 than in referent (unexposed) subjects (Waidyanatha and Rappaport 2001). Hemoglobin adducts were not detected. The correlation between NQ-Ab adducts and workplace naphthalene concentrations was not statistically significant. Given the long half-life of albumin and the measurement of protein adducts after only a single day of exposure, the lack of statistical correlation was not unexpected. The data showed that NQ-Ab may be a marker of long-term exposure to JP-8; however, an estimate of exposure over several days is necessary to validate the use of this marker for chronic exposure to JP-8.

FACTORS THAT MODIFY INTERNAL DOSE OF JP-8

Dermal Absorption

Because JP-8 is a complex mixture of hydrocarbons of widely varying vapor pressure and lipophilicity, uptake by the dermal route is highly affected by the physicochemical properties of the mixture. After dermal application to uncovered skin, individual volatile components may evaporate from the skin surface or penetrate the skin and pass into the venous blood for distribution throughout the body. For substances of low molecular weight, volatile compounds evaporate preferentially. Less-volatile, longer-chain hydrocarbons tend to be hydrophobic (i.e., they have a larger log K_{ow}). Low volatility dictates prolonged skin contact, and the higher hydrophobicity leads to faster dermal penetration. Direct applications of JP-8 to skin produces higher skin and systemic exposures to the higher-molecular-weight components of the distillate mixture.

Several JP-8 additives exhibit high dermal penetration. Using static diffusion cells with excised, denuded F-344 rat skin, Garrett et al. (1999) demonstrated a steady-state flux of 41.4 $\mu g/cm^2$ per hour with the following constituents: diethylene glycol monomethylether (DiEGME) > dodecane > methylnaphthalene > trimethyl benzene > undecane > naphthalene > decane > ethylbenzene > nonane > tridecane. Riviere et al. (1999) studied the systemic absorption and cutaneous distribution of JP-8 across isolated porcine skin flap. Overall percutaneous absorption of dodecane, hexadecane, and naphthalene was inversely related to skin deposition and was influenced by the fuel type and performance additives. Dermal absorption of jet fuel was complicated by the presence of different patterns of binding to the stratum corneum and subcutaneous fat. Riviere et al. (1999) estimated that if both hands were constantly wet with JP-8, about 17 mg of hydrocarbons would penetrate systemically through skin. Baynes et al. (2001) evaluated the influence of JP-8 performance additives on the absorption of JP-8 components. They used porcine skin flaps to characterize chemical-biologic interactions that modulate diffusion of various JP-8 components in vitro. Isolated perfused porcine skin flaps were employed to evaluate diffusion in a model system with viable vasculature. Several performance additives in JP-8 increased the retention of various JP-8 components (e.g., naphthalene and dodecane) on the skin surface. Performance additives, such as DiEGME, may bind to some components of JP-8 and can increase their dermal retention. This could reduce overall systemic absorption of JP-8 fuel during short-term exposure. However, overall systemic absorption may be greater over long periods of continuous dermal contact because of a persistent dermal retention.

Inhalation Exposure

Air Force personnel are most likely to be exposed to JP-8 vapors, although mixed vapor and aerosol exposures may occur during aircraft cold starts or specialized situations such as when the fuel is sprayed as a dust suppressant (CDC 1999). Inhalation of JP-8 vapors or aerosols will result in respiratory tract deposition and systemic absorption of some fuel components.

Inhaled vapors, even those with high volatility and low water solubility, may be largely absorbed in the nasal cavity rather than in the lungs, although significant lung uptake may also occur (Dahl 1989). Following vapor deposition, the extent of systemic absorption depends on the blood:air partition coefficients for the individual fuel components. Monooxygenase-mediated metabolism in the nose, and to a lesser extent in the lung, may activate certain aromatic components of JP-8, possibly resulting in toxic effects at the site of deposition. Repeated inhalation of JP-8 may result in activation of metabolic enzymes in the respiratory tract and, therefore, have an impact on the extent of production of both nontoxic and toxic metabolites. Monooxygenase-mediated metabolism is not expected to have a major effect on JP-8 aliphatic components. Nasal exposures have been implicated as an important source of JP-8 exposure because metabolites of toluene and xylene have been shown to accumulate in nasal mucosa and olfactory bulbs of mice exposed to toluene and xylene by inhalation (Ghantous 1990).

It is generally believed that greater overall deposition of JP-8 might occur following inhalation of respirable aerosols, due to the presence of more high-molecular-weight compounds, compared to that occurring upon vapor inhalation. As with vapors, systemic absorption may occur following deposition. The sites of respiratory tract deposition will vary depending on aerosol particle size and whether oral or nasal breathing occurs. For example, for an aerosol particle with a mass median aerodynamic diameter of 3 mm, approximately 60% will deposit in the nose and 10-15% in the pulmonary region in a nasal breathing human. With oral breathing, approximately 80% will deposit in the pulmonary region. The extent of overall respiratory tract deposition drops dramatically to approximately 40% with particle sizes less than 1 mm.

Oral Exposure

Although of questionable relevance for determining or assessing a PEL for JP-8, oral-toxicity studies with fuels have the advantage of administering the sample without prior fractionation of components. However, other factors related to fractional absorption, oral uptake rates, metabolic clearance, and tissue storage of the individual components lead to differential systemic doses of individual components that might influence toxicity.

ASSESSING DOSIMTERY IN TOXICOLOGY STUDIES

JP-8 is a complex distillate fuel and, therefore, specific exposure scenarios in the workplace, in controlled human studies, and in experimental animal toxicology studies can lead to preferential enhancement of high- or low-volatility components of the fuel. When evaluating human and animal toxicology studies, it is important to know how exposures in workers, volunteers, or animals were generated from liquid JP-8. The process used to generate specific atmospheres of JP-8 will cause enrichment of specific groups of compounds. When describing exposures to JP8, the relationship of various exposure routes and their consequences for altered absorption of low- or high-volatility compounds in the fuel should be specified. This section provides background information on the consequences of different exposure routes in altering absorbed doses of JP-8 fuel components. This information is relevant to interpreting dosing scenarios used in animal toxicity studies and exposures in occupational studies.

Inhalation studies using JP-8 are complicated because the atmosphere may contain a mixture of vapor constituents and liquid-droplet aerosols of fuel. Many of the procedures for atmospheric generation lead to fractionation of the fuel, which may be problematic because there is differential toxicity of fuel components. There are methods of atmosphere generation of vapors that preserve the composition of the original fuel. One method involves heating the fuel in a tube to achieve complete volatilization before introduction of vapor into the experimental chamber atmosphere. This procedure produces vapor atmospheres with a composition similar to that of the fuel. Another method of atmosphere generation involves bubbling air or nitrogen through the fuel and transporting the resulting atmosphere to the exposure chamber. Both of these systems require care to ensure that the final chamber atmosphere does not contain aerosols created by vigorous agitation of the liquid fuel or condensation of the vapors after cooling of the atmosphere. The latter method will generally produce an atmosphere that is enriched with the lower-molecular-weight, more-volatile compounds in the chamber atmosphere compared to the original fuel.

Achieving high concentrations of JP-8 in an exposure chamber requires introduction of a mixture of vapor and aerosol. The size of the aerosol particles determines the lung region where they preferentially deposit. The composition of the aerosol droplets will depend on the methods used for their generation and the "age" and physical characteristics of the aerosol. An aerosol generated from fuel should initially be similar in composition to fuel. As time passes, the preferential volatilization of low-molecular-weight components produces an aerosol enriched in the longer-chain, higher-molecular-weight components. The extent of enrichment depends both on the time that elapses between the production of the aerosol and its presentation in the breathing zone of the subjects and on the residence time of aerosol in the chamber. Aerosol exposure can lead to inhalation of vapors enriched in the low-molecular-weight components and to surface deposition of liquid droplets enriched in higher-molecular-weight *n*-alkanes within the respiratory tract. The above considerations should serve as a guide to the reader as data from human and animal toxicity studies are presented in subsequent chapters.

CONCLUSIONS AND RECOMMENDATIONS

Occupational exposure to JP-8 occurs primarily during transportation and storage of the fuel, aircraft fueling and defueling, maintenance of aircrafts, cold engine starts and performance testing, and operation and maintenance of other Air Force equipment and machinery. The Air Force personnel working in aircraft fuel-cell maintenance shops and fuel-specialty and fuel-transportation shops are probably at greatest risk for exposure to JP-8. Because fuel tanks with cross ventilation have much lower concentrations of JP-8 than fuel tanks with poor ventilation, the subcommittee recommends that the Air Force properly cross-ventilate fuel tanks when personnel are working in them.

The subcommittee concludes that in addition to inhalation exposure, the potential exists for a substantial contribution to overall JP-8 exposure by the dermal route, including mucous membranes and the eyes, either by contact with vapors and aerosols or by direct skin contact with JP-8. Therefore, the subcommittee recommends that the PEL adopted by the Air Force have a skin notation. The subcommittee also recommends that dermal exposures of Air Force personnel to JP-8 be minimized by the use of appropriate protective clothing or other measures. It further recommends that the Air Force evaluate the effectiveness of various protective clothing for personnel who are likely to come into contact with JP-8 and that it use the most effective protective clothing.

Because there is the potential for substantial exposure of troops to JP-8 when it is used to suppress desert sand and as a method of obscuring troops and equipment, the subcommittee recommends that the DOD no longer use JP-8 for those purposes.

The commercial airline jet fuel, Jet A, is similar in composition to JP-8, although JP-8 contains several additives not included in Jet A. Jet A is used widely in commercial aircraft, and the exposed cohort is large; however, the subcommittee did not find epidemiologic studies of this cohort. The subcommittee recommends that an epidemiologic study of commercial airline employees exposed to Jet A be conducted to characterize their exposures and to determine whether any health effects are associated with Jet A exposure.

Because there is evidence that aerosols are more toxic than vapors, the subcommittee recommends development of an analytic method to distinguish between aerosol and vapor exposures in occupational settings.

Finally, the subcommittee recommends that the Air Force monitor the workplace where JP-8 is being used to ensure that exposure to vapors and aerosols does not exceed safe levels.

REFERENCES

ATSDR (Agency for Toxic Substances and Disease Registry). 1998. Toxicological Profile for Jet Fuels (JP-5 and JP-8). U.S. Department of Health and Human Services, Public Health Service, Agency for Toxic Substances and Disease Registry, Atlanta, GA.

Baynes, R.E., J.D. Brooks, K. Budsaba, C.E. Smith, and J.E. Riviere. 2001. Mixture effects of JP-8 additives on the dermal disposition of jet fuel components. Toxicol. Appl. Pharmacol. 175(3):269-281.

Carlton, G.N., and L.B. Smith. 2000. Exposures to jet fuel and benzene during aircraft fuel tank repair in the U.S. Air Force. Appl. Occup. Environ. Hyg. 15(6):485-491.

CDC (Centers for Disease Control and Prevention). 1999. Background Document on Gulf War-Related Research for the Health Impact of Chemical Exposures During the Gulf War: Research Planning Conference. Public Health Service, Bethesda, MD (as cited in Cheng et al. 2001).

Cheng, Y.S., Y. Zhou, J. Chow, J. Watson, and C. Frazier. 2001. Chemical composition of aerosol from kerosene heaters burning jet fuels. Aerosol. Sci. Technol. 35:949-957.

Dahl, A.R. 1989. Metabolic characteristics of the respiratory tract. Pp. 141-160 in Concepts in Inhalation Toxicology, R.O. McClellan, and R.F. Henderson, eds. New York: Hemisphere Publishing Co.

Egeghy, P., and S. Rappaport. 2001. Measurement of benzene and naphthalene in air and breath in the U.S. Air Force as an indicator of JP8 exposure. Pp. 38-54 in JP8 Final Risk Assessment. The Institute of Environmental and Human Health (TIEHH), Lubbock, TX. August 2001.

Garrett, C.M., D.L. Pollard, T.E. Miller, and J.N. McDougal. 1999. In vitro dermal absorption of jet fuel (JP-8) and its components. Conference Topics in Toxicology and Risk Assessment 12-15 April 1999, Wright-Patterson AFB, OH.

Ghantous, H., L. Dencker, J. Gabrielsson, B.R. Danielsson, and K. Bergman. 1990. Accumulation and turnover of metabolites of toluene and xylene in nasal mucosa and olfactory bulb in the mouse. Pharmacol. Toxicol. 66(2):87-92.

Gibson, R., J. Pleil, S. Smith, and D. Toschlog. 2001. Assessment of JP8 in blood. Pp. 29-31 in JP8 Final Risk Assessment. The Institute of Environmental and Human Health (TIEHH), Lubbock, TX. August 2001.

Henz, K. 1998. Survey of Jet Fuels Procured by the Defense Energy Support Center, 1990-1996. Defense Logistics Agencies, Ft. Belvior, VA.

Nylander-French, L.A., and J.D. Archer. 2001. Quantification of dermal exposure to jet fuel, risk assessment of acute exposure to jet fuel. Pp. 25-28 in JP8 Final Risk Assessment. The Institute of Environmental and Human Health (TIEHH), Lubbock, TX. August 2001.

Pleil, J.D. 2001. Direct measurement of total body burden of JP8 jet fuel (breath), risk assessment of acute exposure to jet fuel. Pp. 32-37 in JP8 Final Risk Assessment. The Institute of Environmental and Human Health (TIEHH), Lubbock, TX. August 2001.

Pleil, J.D., L.B. Smith, and S.D. Zelnick. 2000. Personal exposure to JP-8 jet fuel vapors and exhaust at air force bases. Environ. Health Perspect. 108(3):183-192.

Riviere, J.E., J.D. Brooks, N.A. Monteiro-Riviere, K. Budsaba, and C.E. Smith. 1999. Dermal absorption and distribution of topically dosed jet fuels jet-A, JP-8, and JP-8(100). Toxicol. Appl. Pharmacol. 160(1):60-75.

Serdar, B., P.P. Egeghy, and S.M. Rappaport. 2001. Urinary benzene, naphthalene, 1- and 2-hydroxynaphthalene as biomarkers of acute (short-term) exposure to JP8. Pp. 57-64 in JP8 Final Risk Assessment. The Institute of Environmental and Human Health (TIEHH), Lubbock, TX. August 2001.

TIEHH (The Institute of Environmental and Human Health). 2001. JP8 Final Risk Assessment. The Institute of Environmental and Human Health, Lubbock, TX. August 2001. 179pp.

Ullrich, S.E. 1999. Dermal application of JP-8 fuel induces immune suppression. Toxicol. Sci. 52(1):61-67.

Waidyanatha, S., and S.M. Rappapot. 2001. Protein adducts as biomarkers of exposure to jet fuel. Pp. 121-178 in JP8 Final Risk Assessment. The Institute of Environmental and Human Health (TIEHH), Lubbock, TX. August 2001.

Witten, M. 2002. JP-8 jet fuel: An overview. [Online] Available: http://www.jp8.org [accessed October 1, 2002].

Wolfe, R.E., E.R. Kinead, M.L. Feldmann, H.F. Leahy, and W.W. Jederberg. 1997. Acute Toxicity Evaluation of JP-8 Jet Fuel and JP-8 Jet Fuel Containing Additives. Govt. Rep. Announce Index (GRA&I) Issue 09.

Yeung, P., A. Rogers, and B. Davies. 1997. Safe working in aircraft fuel tanks: An Australian experience. Appl. Occup. Environ. Hyg. 12(9):587-594.

Zhou, Y., and Y.S. Cheng. 2000. Characterization of emission from kerosine heaters in an unvented tent. Aerosol Sci. Technol. 33:510-524.

3

Toxicokinetics and Toxicodynamics of Jet-Propulsion Fuel 8

This chapter contains a review of toxicokinetics and toxicodynamics data for jet-propulsion fuel 8 (JP-8). Given the presence of hundreds of hydrocarbons in JP-8, it is impractical to describe here the toxicokinetics and disposition of each component hydrocarbon. In the previous National Research Council (NRC) report, *Permissible Exposure Levels for Selected Military Fuel Vapors*, the toxicokinetics of some toxic components of JP-8—including benzene and alkylbenzenes (such as xylenes and toluene)—were discussed in detail (NRC 1996).

The following general principles were applied to describe the toxicokinetics of JP-8. The major determinants of hydrocarbon toxicokinetics following systemic uptake are disposition-related physiologic measures, such as alveolar ventilation, cardiac output and blood flow to organs, partition coefficients, and organ volume. Hydrocarbons with high blood:air partition coefficients will be absorbed to a greater extent than chemicals with poor blood solubility. Given that most hydrocarbons have fairly high fat:air and fat:blood partition coefficients, it is not surprising that fat or adipose tissue is a major storage depot for many of these JP-8 components. For hydrocarbons with high fat:blood partition coefficients, metabolic clearance following cessation of exposure is more important than that during exposure. Given the accumulation of hydrocarbons or their metabolites in lipid-rich tissues, the absence of hydrocarbons and their metabolites in exhaled air, blood, or urine does not necessarily mean the absence of systemic exposure. With regard to metabolism, cytochrome P450 (CYP450) enzymes metabolize most hydrocarbons by such reactions as aliphatic hydroxylation, aromatic hydroxylation, and epoxidation. For many hydrocarbons, alcohol and aldehyde dehydrogenases play an important role in metabolizing alcohols to their corresponding keto acids. Phase II reactions, including conjugation with glutathione, glucuronic acid, sulfate, and glycine, are important in formation of water-soluble metabolites. The following discussion of toxicokinetics will be limited to a brief summary of disposition of only toxicologically relevant components of JP-8.

BENZENE

Benzene is a minor component of JP-8 (<1%), but its high volatility, its flammability, and its moderate water solubility make it an important component of JP-8 exposure (ACGIH 1996; Paustenbach 2000). Benzene is a potent genotoxicant and a recognized human carcinogen. Dose-dependent bone-marrow suppression, pancytopenia (e.g., aplastic anemia), and neurologic toxicities can occur after high-dose benzene exposure (Evans et al. 1981; McConnell 1993). The metabolism of benzene has been discussed in a previous NRC report (NRC 1996) and by ACGIH (1996). Benzene is metabolized primarily via the hepatic CYP450 system to benzene oxide, which is biotransformed to 1,2-dihydrodiol, which leads to catechol formation. Benzene oxide can also rearrange nonenzymatically to phenol, which is biotransformed to hydroquinone and benzoquinone. The water-soluble metabolites of benzene (phase II conjugative metabolism) are readily excreted (Paustenbach et al. 1993). Combined exposure to catechol and hydroquinone metabolites has been implicated in benzene's genotoxicity (Robertson et al. 1991). Benzene and its metabolites have been shown to accumulate in humans (e.g., they appear in exhaled air and urine) after repeated exposure to benzene.

ALKYLBENZENES

The alkylbenzenes (single-ring aromatic compounds with single or multiple aliphatic side chains) are constituents of JP-8. Toluene (methylbenzene) and mixed xylenes (*o*-, *m*-, and *p*-) are present in JP-8 and have been identified as potential neurotoxic chemicals after sufficiently high intentional, accidental, or occupational exposures (Gamberale and Hultengren 1972; Boor and Hurtig 1977; Klaucke et al. 1982; Hipolito 1980).

The toxicokinetics of toluene have been well characterized in laboratory animals (Benignus et al. 1981; Benignus 1982). In rats, after inhalation exposure, toluene was quickly absorbed and distributed in lipoidal and highly vascularized tissues. Within an hour of inhalation exposure at 2,167 mg/m^3, about 95% of maximal concentrations in blood and brain were achieved. Toluene is metabolized principally by series of oxidation reactions that lead to benzoic acid, which is conjugated with glycine to form hippuric acid. Unchanged toluene is readily removed in exhaled air.

Xylene vapor is rapidly absorbed from the lungs, and xylene liquid and vapor are absorbed slowly through skin. More than 90% of the absorbed xylene is metabolized to methylhippuric acid, which is readily excreted in urine. On repeated administration, xylene auto-induces some P450 enzymes, which metabolize the methyl side chain to toluic acid (methylbenzoic acid), which is also rapidly excreted. Additional metabolites of xylenes are dimethylphenol and methylbenzyl alcohol (Langman 1994).

Exposures to toluene appear to have an initial central nervous system (CNS) stimulatory effect; intentional or accidental high exposures are associated with CNS depression (Gamberale and Hultengren 1972; Boor and Hurtig 1977). In humans, accidental exposures to xylene at up to 10,000 ppm resulted in epileptic seizure, complete amnesia, cerebral hemorrhage, unconsciousness, and ventricular fibrillation (Low et al. 1989). There are no relevant studies on low-level chronic exposure to toluene or mixed xylenes in JP-8.

C9-C13 ALIPHATIC AND AROMATIC HYDROCARBONS

Long-chain and branched hydrocarbons that are primary components of JP-8, include *n*-nonane, *n*-decane, *n*-dodecane, *n*-tridecane, isopropylbenzene, *n*-propylbenzene, trimethylbenzene, *n*-dimethylbenzene, naphthalene, *n*-pentylbenzene, and *n*-triethylbenzene. Inhaled long-chain aliphatic hydrocarbons generally show poor blood uptake because of lower blood solubility. They have relatively high lipid:blood partition coefficients; this can result in accumulation in lipid-rich tissues, such as brain and fat. In laboratory studies, brain concentrations of hydrocarbons and their metabolites greatly exceed their plasma concentrations.

Zahlsen et al. (1993) found that C8-C10 hydrocarbons were extremely well absorbed and their tissue distribution in brain and fat were largely dependent on the number of carbon atoms. *n*-Nonane is one component of JP-8 and it is metabolized at relatively high rates to hydroxyl derivatives, which are converted to the corresponding ketone. Other important hydrocarbons (from a quantitative perspective) are decane, dodecane, tetradecane, and hexadecane; they are each present at concentrations of over 10% in the liquid form of JP-8. Decane has immunotoxic potential, and its lipophilicity is similar to that of hexane and octane. Decane metabolism is similar to that of nonane: it is metabolized to the corresponding ketone though an intermediate hydroxylation step. Dodecane, tetradecane, and hexadecane are likely to be metabolized similarly, but their metabolism is not well characterized. Microorganisms also metabolize these hydrocarbons extensively. Because aliphatic fractions above C13 have relatively low volatility, they are unlikely to be present at toxicologically significant concentrations in JP-8 vapor (Sandmeyer 1981). The presence of the very-long-chain aliphatics (i.e., above C13) in JP-8 aerosol is not known.

Sufficiently high exposures to alkylbenzenes—such as *n*-diethylbenzene, *n*-triethylbenzene, *n*-trimethylbenzene, and isopropylbenzene—can produce adverse motor and sensory effects in rats after inhalation. Gagnaire et al. (1990) reported decreased motor and sensory conduction velocities and decreased amplitude of the sensory action potential of the tail nerve in rats exposed repeatedly to a mixture of diethylbenzene and its major metabolite, 1,2-diacetylbenzene (DAB). In the rat, 1,2-diethylbenzene was about 5 times more potent a neurotoxicant than *n*-hexane (Gagnaire et al. 1990). Spinal-cord axonal swelling with partial demyelination has been associated with DAB exposure in rats (Kim et al. 2001). The neurotoxicity of 1,2-diethylbenzene appears to be related to formation of protein complexes: reaction with amino acids of proteins to form pyrolated polymers that lead to protein cross links.

PHYSIOLOGICALLY BASED PHARMACOKINETIC MODELS OF BENZENE, NONANE, AND C9-C12 OR C9-C17 ALIPHATIC HYDROCARBONS

Physiologically based pharmacokinetic (PB-PK) models for some JP-8 components have been developed to understand the relationship between vapor concentrations and accumulation in tissue and blood compartments. When appropriately developed and validated, PB-PK models can provide a time course of distribution of a chemical or its metabolites in tissues and show the effect of changing physiologic characteristics on plasma and tissue concentrations. PB-PK models have been applied to predict toxicokinetic parameters and to scale dose in different species.

Kinetic studies have been done with representative compounds, including benzene, short-chain alkanes and iso-alkanes, and naphthalene. There is kinetic information about the higher-molecular-weight compounds that become less volatile and less water- and blood-soluble and more lipid-soluble. PK studies provide information on those components of JP-8, which might assist in biomonitoring for components that are found in JP-8 at higher percentages and in beginning to assess their metabolites.

PB-PK models have been applied to evaluate the behavior of benzene and alkylbenzenes alone and in combination (Medinsky et al. 1989, 1995; Purcell et al. 1990). Some of the published models examine human populations exposed at concentrations found in the environment (Sherwood and Sinclair 1999) or at occupationally relevant concentrations in volunteers (Beningnus et al. 1998). The models all use flow-limited uptake of chemicals by tissues and metabolic clearance from the liver compartment. Interaction models rely on competition of the various substrates for enzymatic clearance in liver (Haddad et al. 2000). Another similarity in all the kinetic models is the representation of the relationship of circulating concentrations of those aromatic hydrocarbons to exhaled air. In the model, all compounds in venous blood are available for gas exchange in the lungs; this leads to simple relationships in which concentration in exhaled air is a straightforward function of concentration in blood, blood:air partition coefficient, total cardiac output, and alveolar ventilation. The models have proved successful for low-molecular-weight hydrocarbons; their application has not been established for the longer-chain *n*-alkanes.

Robinson (2000) described a PB-PK model of JP-8 constituents that used nonane as a marker. Nonane has been considered a surrogate biomarker for JP-8 aliphatic hydrocarbons (e.g., C9-C12 or C9-C17 aliphatic hydrocarbons) in breath, and it distributes preferentially to brain. Nonane disposition was described in a PB-PK model that includes its distribution in blood, lungs, liver, muscle, and fat. The model was developed with "inhouse" F-344 rat inhalation data, including blood concentrations, and was validated with published data on Sprague-Dawley rats. The model was used to predict the body burden of nonane after known occupational JP-8 exposures. There was generally good agreement between the PB-PK model based on inhouse data and published data. Overall, blood and brain nonane concentrations were well predicted over a 10-fold range of concentrations in inhaled air. Limitations of the model include the lack of empirical data on metabolic measures such as Vmax and Km; the information on metabolic rates (hydroxylation followed by metabolism to corresponding keto form); and the lack of data on alveolar ventilation rates. Given those limitations, the model overpredicted blood concentrations and underpredicted the slope of terminal elimination. The incorrect predictions appear related to oversimplifications of several tissue compartments and the use of only a small number of homogeneous compartments. Efforts were made to extrapolate the rat PB-PK model of nonane to predict results of exposures of fuel-tank entry workers and attendants. The rat PB-PK model overpredicted the blood concentrations of nonane determined in a JP-8 acute-exposure human study (Pleil et al. 2000), possibly because of the higher affinity of nonane to rat red blood cells than to human red blood cells and the lack of information on human metabolic measures and partition coefficients.

Dixon et al. (2001) described a preliminary PB-PK model to predict JP-8 concentrations in Air Force fuel-cell maintenance workers. The model used data from PB-PK models of naphthalene inhalation in mice and rats and nonane inhalation in rats. In addition to inhalation, a pathway of dermal exposure and a skin compartment were included. For highly exposed people, the PB-PK model was generally in agreement with exhaled-air naphthalene concentrations; however, predictions for the low-exposure scenarios were grossly underestimated, especially in female workers, by a factor of 10. The model did not predict blood and urinary concentrations. The major limitation of the Dixon et al. (2001) study was the lack of appropriate human data (e.g., metabolic measures, blood and tissue partition coefficients, and diffusion rates). The Dixon et al. (2001) model predicted a rapid decline in naphthalene concentrations in all compartments after exposure except liver, fat, and brain. The model predicted accumulation in liver, brain, and fat tissues for a 7-day period that included 4-hr exposures on 5 days. Competition for enzyme does not occur only from interactions of different inhaled compounds. Interactions can also occur between inhaled compounds and metabolites formed in the body that require similar enzymes for biotransformation. Detailed kinetic studies with both benzene and *n*-hexane show inhibition of later metabolic steps, phenol to hydroquinone or methyl *n*-butyl ketone to 2,5-hexane dione, by high concentrations of inhaled benzene or hexane, respectively (Medinsky et al. 1989; Andersen and Clewell 1984).

Recent work with another highly lipophilic compound with low blood:air partitioning and high fat solubility may be instructive for developing predictive PB-PK models for higher-molecular-weight *n*-alkanes found in JP-8 (Andersen et al. 2001). Octamethylcyclotetrasiloxane (D4) has a Pb value of about 2.0 and a fat:blood partition coefficient of 500-600. A set of detailed studies was conducted to measure behavior of D4 in blood, exhaled air, and tissues during and after 6-hr exposures in rats. The data could not be described with a conventional PB-PK model, because there were discrepancies between blood and exhaled-air concentrations. By focusing on the exhaled-air concentrations as a measure of free D4 in plasma, a PB-PK model was developed that included a pool of blood D4 that was not available for exhalation and was probably sequestered in blood lipids (Andersen et al. 2001). The ability to discern deviations from conventional models and a model with sequestered blood D4 depended on the availability of a robust, quality-controlled study with multiple collections of blood and exhaled-air concentrations. No such data are available on any of the long-chain alkanes found in JP-8 in laboratory animals or human volunteers.

Overall, almost all recently developed PB-PK models of various JP-8 components based on nonane or naphthalene do not predict accurately the JP-8 or component concentrations in blood, breath, or tissue. Given that most JP-8 components have a propensity to accumulate in lipid-rich tissues, there is a need for more accurate PB-PK models that can predict tissue concentrations of important classes of JP-8 components. Such data are critical in assessing the long-term safety of JP-8 occupational exposure.

TOXICOKINETICS-RELATED INTERACTIONS AMONG HYDROCARBON FUEL COMPONENTS

With over 200 hydrocarbons present in JP-8, there is the possibility of toxicokinetics- and toxicodynamics-related antagonistic, additive, and synergistic interactions among various hydrocarbon components. The toxicokinetic parameters of individual chemicals in a complex mixture such as JP-8 are very different from the toxicokinetic profiles of individual chemicals. Chemical-chemical interactions may be related to mutual induction of competing metabolic and elimination pathways or mutual inhibition of absorption, distribution, metabolism, and excretion. Similar CYP450 enzymes and phase II conjugative metabolic pathways metabolize many alkane hydrocarbons (NRC 1996).

Depending on the K_m and V_{max} for the metabolism of individual chemicals, competitive metabolic interactions can result in lower or higher concentrations of chemicals and their metabolites. For example, toluene inhibits the oxidative metabolism of benzene and reduces blood and bone marrow toxicity of benzene (Purcell et al. 1990). Tardif et al. (1992) evaluated dose-dependent interactions between toluene and xylene. The combined exposure to toluene and xylene resulted in lower amounts of urinary hippuric acid (20-30%) and methylhippuric acid (4-40%) than exposure to the individual agents. The greatest reductions were observed in the group exposed to both toluene and xylene at 150 ppm. The blood and brain concentrations of both toluene and xylene were 230-500% higher than the concentrations following equivalent exposures to the individual chemicals alone. A similar interaction was observed after repeated exposure. Those interactions were probably related to competitive and mutual inhibition of oxidative and conjugative metabolism by both xylene and toluene inasmuch as the two compounds use similar metabolic pathways.

Morata et al. (1995) reported metabolic and toxic interactions between toluene and several components of jet fuels. In rats exposed to toluene and hexane for 18 hr/day for 61 days, a synergistic reduction in auditory sensitivity occurred in the toluene-plus-hexane group that persisted for 365 days. Pryor and Rebert (1992) evaluated possible interactions between toluene and *n*-hexane exposures (individual vs. mixture protocols). Generally, the addition of toluene to the *n*-hexane exposure reduced *n*-hexane-induced neuropathy. Perbellini et al. (1992) showed that inhibition of *n*-hexane-induced neuropathy by toluene was related to inhibition of *n*-hexane metabolism to its neurotoxic metabolite, 2,5-hexanedione. *n*-Hexane and toluene use similar oxidative pathways for their primary metabolism.

Although most chemical-chemical interactions related to JP-8 tend to result from competitive inhibition of oxidative metabolism, induction of metabolism can result in increased toxicity of some components of JP-8, such as benzene, hexane, and naphthalene. Dosing et al. (1985, 1988) reported increased oxidative metabolism (antipyrene clearance) in personnel exposed to the jet fuel by vapor inhalation. The effect of metabolic induction by JP-8 hydrocarbons on the toxicity of chemicals that produce toxicity through metabolite(s) has not been evaluated. Ethanol is an inducer of hepatic CYP2E, which also metabolizes benzene. Ethanol ingestion might lead to increased metabolic activation of benzene, which might lead to bone marrow toxicity, hematotoxicity, and possibly leukemia (ACGIH 1996). The metabolic and toxicologic interactions between ethanol consumption and the various hydrocarbon components in JP-8 have not been carefully studied.

Coupling of validated PB-PK models with a better grasp of the constituents or metabolites associated with target organ toxicity makes it possible to assess the effects of phenotypic variants and other forms of variation on tissue doses of inhaled compounds and their metabolites. Such assessments are becoming routine, but in the absence of a model for calculating the consequences of the variants in relation to tissue dose, the presence of the phenotypic variations is impossible to assess quantitatively in a rigorous manner.

TOXICOKINETIC-RELATED INDIVIDUAL SUSCEPTIBILITY FACTORS

Toxicokinetic-related individual susceptibility factors may include individual differences in rates of absorption, concentration at target sites, metabolic activation, and detoxification of individual chemicals. In addition, people can differ in the induction of adaptive protective responses. Genetics, pregnancy, lifestyle factors (such as smoking, alcohol-drinking, and recreational drug abuse), age, health status, ethnicity, and prior and concurrent occupational exposures to multiple chemicals contribute to a person's susceptibility to chemical toxicity.

Genetic polymorphism in drug-metabolizing enzymes has been shown to be a major contributing factor of individual susceptibility to chemical toxicity (Wiencke et al. 1997; Snyder and Hedli 1996). Of particular interest is a recent, preliminary study (Butler et al. 2001; Frame and Dickerson 2001) that evaluated allelic variants in three polymorphic genes: CYP2E1 (for activation of benzene), GST M1 (for inactivation of bioactivated metabolites), and NAD(P)H quinone reductase (NQO1, involved in inactivation of benzoquinones). A minor DraI allele C in the CYP2E1 gene contributing to high activity is present in about 10-14% of the population and was correlated with formation of DNA adducts after low-level exposure (Kato et al. 1995). The C-to-T transition at base pair 609 of exon 6 in the NQO1 gene leads to about a 3-fold decrease in activity and is present in about 50% of the population (Rothman et al. 1997; Wiencke et al. 1997). Because of a homozygous deletion of GST T1 gene, about 30% of the population does not express an active GST T1 enzyme (Ketterer et al. 1992). Overall, this preliminary study did not identify any important influence of genetic polymorphism in the above three genes on individual toxicokinetics (body burden of JP-8) or on adverse effects following brief exposure to JP-8 (Butler et al. 2001; Frame and Dickerson 2001). In a study by Soderkvist et al. (1996), a significant correlation was found with the GST M1 null genotype and the risk of chronic toxic encephalopathy in patients with a history of long-term exposure to industrial hydrocarbon solvents. GST M1 is a polymorphic enzyme; about 50% of the population expresses a null GST M1 genotype. Frame and Dickerson (2001) found a significant increase in GST M1 null phenotype distribution in naphthalene-exposed people (60.6% GST 1 null phenotype) compared with unexposed people (45.4% GST null phenotype). Despite that trend, there was no relationship between GST M1 null distribution and high or low exposure to naphthalene on the basis of exhaled-air concentration. That suggests that environmental and lifestyle factors contributed to an increase in GST M1 null genotype independently of JP-8 exposure. The presence of GST M1 null genotype after low or moderate exposure may predispose people to adverse health effects compared to people with normal GST M1 activities.

REFERENCES

ACGIH (American Conference of Governmental Industrial Hygienists). 1996. Threshold Limit Values for Chemical Substances and Physical Agents and Biological Exposure Indices. American Conference of Governmental Industrial Hygienists, Cincinnati, OH.

Andersen, M.E., and H.J. Clewell III. 1984. Pharmacokinetic interactions of mixtures. Pp. 226-238 in Proceedings of the 14th Annual Conference on Environmental Toxicology, 15-17 Nov. 1983, Dayton, OH. AFAMRL-TR-83-099. AD-A146 400. Air Force Aerospace Medical Division, Wright-Patterson AFB, OH. August.

Andersen, M.E., R. Sarangapani, R.H. Reitz, R.H. Gallavan, I.D. Dobrev, and K.P. Plotzke. 2001. Physiological modeling reveals novel pharmacokinetic behavior of inhaled octamethylcyclotetrasiloxane in rats. Toxicol. Sci. 60(2):214-231.

Benignus, V.A., K.E. Mueller, C.N. Barton, and J.A. Bittikoffer. 1981. Toluene levels in blood and brain of rats during and after respiratory exposure. Toxicol. Appl. Pharmacol. 61(3):326-334.

Benignus, V.A. 1982. Neurobehavioral effects of toluene: A review. Neurobehav. Toxicol. Teratol. 3(4):407-416.

Benignus, V.A., W.K. Boyes, and P.J. Bushnell. 1998. A dosimetric analysis of behavioral effects of acute toluene exposure in rats and humans. Toxicol. Sci. 43(2):186-195.

Boor, J.W., and H.I. Hurtig. 1977. Persistent cerebellar ataxia after exposure to toluene. Ann. Neurol. 2(5):440-442.

Butler, M.A., C.A. Flugel, E.F. Krieg, J.E. Snawder, and J.S. Kesner. 2001. Gene-environment interactions and exposure to JP8 jet fuel. Pp. 76-80 in JP8 Final Risk Assessment. The Institute of Environmental and Human Health (TIEHH), Lubbock, TX. August 2001.

Dixon, K.R, E.P. Albers, and C. Chappell. 2001. A model for predicting health risk to exposure to JP8 jet fuel. Pp. 140-151 in JP8 Final Risk Assessment. The Institute of Environmental and Human Health (TIEHH), Lubbock, TX. August 2001.

Dosing, M., S. Loft, and E. Schroeder. 1985. Jet fuel and liver function. Scand. J. Work Environ. Health. 11(6):433-437.

Dosing, M., S. Loft, J. Sonne, and E. Schroeder. 1988. Antipyrene and metronidazole metabolism during occupational exposure to gasoline. Int. Arch. Occup. Environ. Health. 60(2):115-118.

Evans, H.L., A.M. Dempster, and C.A. Snyder. 1981. Behavioral changes in mice following benzene inhalation. Neurobehav. Toxicol. Teratol. 3(4):481-485.

Frame, L.T., and R.L. Dickerson. 2001. The human glutathione-S-transferase M1 (GSTM1) polymorphism as a risk factor for acute toxicity from jet fuel exposure. Pp. 87-90 in JP8 Final Risk Assessment. The Institute of Environmental and Human Health (TIEHH), Lubbock, TX. August 2001.

Gagnaire, F., B. Marignac, and J. de Ceaurriz. 1990. Diethylbenzene-induced sensorimotor neuropathy in rats. J. Appl. Toxicol. 10(2):105-112.

Gamberale, F., and M. Hultengren. 1972. Toluene exposure. II. Psychophysiological functions. Work Environ. Health 9(3):131-139.

Haddad, S., G. Charest-Tardif, and K. Krishnan. 2000. Physiologically based modeling of the maximal effect of metabolic interactions on the kinetics of components of complex chemical mixtures. J. Toxicol. Environ. Health A 61(3):209-223.

Hipolito, R.N. 1980. Xylene poisoning in laboratory workers: Case reports and discussion. Lab. Med. 11:593-595.

Kato, S., E.D. Bowman, A.M. Harrington, B. Blomeke, and P.G. Shields. 1995. Human lung carcinogen-DNA adduct levels mediated by genetic polymorphism in vivo. J. Nat. Can. Inst. 87(12):902-907.

Ketterer, B., J.M. Harris, G. Talaska, D.J. Meyer, S.E. Pemble, J.B. Taylor, N.P. Lang, and F.F. Kadulbar. 1992. The human glutathione S-transferase supergene family, its polymorphism, and its effect on susceptibility to lung cancer. Environ. Health. Perspect. 98:87-94.

Kim, M.S. , M.I. Sabri, V.H. Miller, R.J. Kayton, D.A. Dixon, and P.S. Spencer. 2001. 1,2-diacetylbenzene, the neurotoxic metabolite of a chromogenic aromatic solvent, induces proximal axonopathy. Toxicol. Appl. Toxicol. 177(2):121-131.

Klaucke, D.N., M. Johansen, and R.L. Vogt. 1982. An outbreak of xylene intoxication in a hospital. Am. J. Ind. Med. 3(2):173-178.

Langman, J.M. 1994. Xylene: Its toxicity, measurement of exposure levels, absorption, metabolism and clearance. Pathology 26(3):301-309.

Low, L.K., J.R. Meeks, and C.R. Mackerer. 1989. Health effects of the alkylbenzenes. II. Xylenes. Toxicol. Ind. Health 5(1):85-105.

McConnell, E.E. 1993. Benzene. Environmental Health Criteria 150. Geneva: World Health Organization.

Medinsky, M.A., E.M. Kenyon, and P.M. Schlosser. 1995. Benzene: A case study in parent chemical and metabolite interactions. Toxicology 105(2-3):225-233.

Medinsky, M.A., P.J. Sabourin, G. Lucier, L.S. Birnbaum, and R.F. Henderson. 1989. A physiological model for simulation of benzene metabolism by rats and mice. Toxicol. Appl. Pharmacol. 99(2):193-206.

Morata, T.C., P. Nylen, A.C. Johnson, and D.E. Dunn. 1995. Auditory and vestibular functions after single or combined exposure to toluene: A review. Arch. Toxicol. 69(7):431-443.

NRC (National Research Council). 1996. Permissible Exposure Levels for Selected Military Fuel Vapors. Washington, DC: National Academy Press.

Paustenbach, D.J. 2000. The practice of exposure assessment: A state-of-the-art review. J. Toxicol. Environ. Health B Crit. Rev. 3(3):179-291.

Paustenbach, D.J., R.D. Bass, and P. Price. 1993. Benzene toxicity and risk assessment, 1972-1992: Implications for future regulation. Environ Health Perspect. 101(Suppl.6):177-200.

Perbellini, L., R. Leone, M.E. Fracasso, F. Brugnone, and M.S. Venturini. 1992. Metabolic interactions between n-hexane and toluene in vivo and in vitro. Int. Arch. Occup. Environ. Health 50(4):351-358.

Pleil, J.D., L.B. Smith, and S.D. Zelnick. 2000. Personal exposure to JP-8 jet fuel vapors and exhaust at air force bases. Environ. Health Perspect. 108(3):183-192.

Pryor, G.T., and C.S. Rebert. 1992. Interactive effects of toluene and hexane on behavior and neurophysiologic responses in Fischer-344 rats. Neurotoxicology 13(1):225-234.

Purcell, K.J., G.H. Carson, M.L. Gargas, M.E. Andersen, and C.C. Travis. 1990. In vivo metabolic interactions of benzene and toluene. Toxicol. Lett. 52(2):141-152.

Robertson, M.L., D.A. Eastmond, and M.T. Smith. 1991. Two benzene metabolites, catechol and hydroquinone, produce a synergistic induction of micronuclei and toxicity in cultured human lymphocytes. Mutat. Res. 249(1):201-209.

Robinson, P.J. 2000. Pharmacokinetic Modeling of JP-8 Jet Fuel Components. I. Nonane and C9-C12 Aliphatic Components. AFRL-HE-WP-TR-2000-0046. U.S. Air Force Research Laboratory, Wright Patterson AFB OH.

Rothman, N., M.T. Smith, R.B. Hayes, R.D. Traver, B. Hoener, S. Campleman, G.L. Li, M. Dosemeci, M. Linet, L. Zhang, L. Xi, S. Wacholder, W. Lu, K.B. Meyer, N. Titenko-Holland, J.T. Stewart, S. Yin, and D. Ross. 1997. Benzene poisoning, a risk factor for hematological malignancy, is associated with the NQO1609C->T mutation and rapid fractional excretion of chlorzoxazone. Cancer. Res. 57(14):2839-2842.

Sandmeyer, E.E. 1981. Aliphatic hydrocarbons. Pp. 3175-3220 in Patty's Industrial Hygiene and Toxicology, Vol. IIB, 3rd Rev. Ed., G.D. Clayton, and F.E. Clayton, eds. New York: John Wiley & Sohn.

Sherwood, R.J., and G.C. Sinclair. 1999. New PBPK model applied to old occupational exposure to benzene. Am. Ind. Hyg. Assoc. J. 60(2)259-265.

Snyder, R., and C.C. Hedli. 1996. An overview of benzene metabolism. Environ Health Perspect. 104(Suppl.6):1165-1171.

Soderkvist, P., A. Ahmadi, A. Akerback, O. Axelson, and U. Flodin. 1996. Glutathione S-transferase M1 null genotype as a risk modifier for solvent-induced chronic toxic encephalopathy. Scand. J. Work Environ. Health 22(5):360-363.

Tardif, R., G.L. Plaa, and J. Brodeur. 1992. Influence of various mixtures of inhaled toluene and xylene on biological monitoring of exposure to these solvents in rats. Can. J. Physiol. Pharmacol. 70(3):385-393.

Wiencke, J.K., M.R. Spitz, A. McMillan, and K.T. Kelsey. 1997. Lung cancer in Mexican-Americans and African-Americans is associated with the wild-type genotype of the NAD(P)H: Quinone oxidoreductase polymorohism. Cancer Epidemiol. Biomarkers Prev. 6(2):87-92.

Zahlsen, K., I. Eide, A.M. Nilsen, and O.G. Nilsen. 1993. Inhalation kinetics of C-8 to C-10 1-alkenes and iso-alkanes in the rat after repeated exposures. Pharmacol. Toxicol. 73(3):163-168.

4

Effects of Jet-Propulsion Fuel 8 on the Respiratory Tract

In this chapter, the subcommittee reviews studies in humans and experimental animals that examined potential respiratory tract effects of jet-propulsion fuel 8 (JP-8), related fuels, and kerosene. The subcommittee uses that information to assess the potential respiratory toxicity of JP-8 in humans. The National Research Council report *Permissible Exposure Levels for Selected Military Fuel Vapors* (NRC 1996) did not include a review of the effects of JP-8 on the respiratory tract.

EFFECTS OF EXPOSURE TO JET FUELS AND KEROSENE IN HUMANS

Few studies have directly or systematically addressed the potential for adverse effects of JP-8 or other jet fuels on the human respiratory tract. Available studies of respiratory tract toxicity of jet fuels and kerosene are described below and summarized in Table 4-1.

Tunnicliffe et al. (1999) reported the effect of occupational exposure to aircraft fuel (type not specified) and jet-stream exhaust on pulmonary function and respiratory symptoms in airport workers. Two hundred twenty-two full-time airport employees were divided into groups with essentially no exposure

TABLE 4-1 Effects of Jet Fuel Exposure on the Respiratory Tract in Humans

Exposure Concentration	Exposure Duration	Results	Reference
Not reported	222 full-time airport employees divided into three exposure groups; high-exposure group, 56 participants exposed for most of day; moderate-exposure group, 83 participants exposed for 1 hr/day; no-exposure group, 86 participants	Adjusted odds ratios for cough with phlegm (3.5) and for runny nose (2.9) significantly associated with frequent exposure; adjusted odds ratios for symptoms of watering eyes, stuffy nose, wheezing, shortness of breath not significant	Tunnicliffe et al. 1999
Exposed group (5,706) had potential occupational exposure to JP-8; control group (5,706) did not work in occupations in which exposure to JP-8 would occur; all subjects were active duty members of US Air Force	Not reported	Analysis of medical records showed that subjects in all groups had similar health-care visit rates; no differences among groups in respiratory illnesses	Gibson et al. 2001a[a]
Measurements taken in breathing zones of subjects; median concentration of naphthalene, 1.9 $\mu g/m^3$ (low-exposure group), 10.4 $\mu g/m^3$ (moderate-exposure group), 447 $\mu g/m^3$ (high-exposure group); median concentration of benzene, 3.1 $\mu g/m^3$ (low-exposure group), 7.45 $\mu g/m^3$ (moderate-exposure	High- and moderate-exposure groups had persistent exposure to JP-8 (defined as at least 1 hr twice per wk for at least 9 mo); low-exposure group had no significant exposure to jet fuel or solvents	Analysis of self-assessment questionnaire did not report differences among groups in measures related to respiratory tract, such as difficulty in breathing	Gibson et al. 2001b[a]

group), 242 $\mu g/m^3$ (high-exposure group)			
Overall average concentration 300 mg/m^3 (range, 85-974 mg/m^3)	Average employment duration, 17 yr	Undefined respiratory tract symptoms, palpitations, and feeling of pressure in chest may have been associated with exposure	Knave et al. 1978
Three families (six adults, three children) exposed to kerosene in their homes at 5.6-79.7 mg/m^3	4-8 mo as result of spill near their homes; exposure estimated at 100 hr/wk	Three children, one adult developed asthma that persisted for more than 2 yr; other adults developed other respiratory tract symptoms, such as sore throat, cough, watery eyes, stuffed noses, chest tightness	Todd and Buick 2000

[a]Additional background information about these studies can be found in Appendix B.

(86 participants), occasional exposure (83 participants exposed about 1 hr/day), and frequent exposure (56 participants exposed for "most of the day"). Participants filled out a self-administered questionnaire regarding respiratory symptoms, were skin-tested for allergies, and underwent spirometry to assess pulmonary function. Because of differences in demographic makeup of the exposure groups (low representation of females and socioeconomic stratification between the unexposed and exposed groups), logistic regression analyses on study data were performed only to compare men in the occasional- and frequent-exposure groups. Analyses were corrected for age and smoking habits and, where appropriate, for the presence of self-reported hay fever or asthma. The adjusted odds ratios (ORs) for cough with phlegm (3.5) and for runny nose (2.9) were significantly associated with frequent jet exhaust exposure. ORs for symptoms of watering eyes, stuffy nose, wheezing, and shortness of breath were not significantly different between the frequently and occasionally exposed groups. The investigators stated that increased adjusted ORs in the frequent-exposure group likely reflected a true association between symptoms and occupational settings, although bias and residual confounding could not be discounted. The investigators also stated that the symptoms suggested exposure to a respiratory irritant; the effects were more closely related to exposure to jet exhaust than to exposure to jet fuel. The Tunnicliffe et al. (1999) study is limited by lack of quantitative exposure assessment, elimination of evaluation of the unexposed group, limited end-point evaluation, lack of correction for subject bias, and the relatively small number of participants. The hypothesis that symptoms of respiratory irritation were due more to jet exhaust than to fuel should be taken with caution; more-recent studies have determined that JP-8 vapor can cause upper respiratory tract irritation in mice (U.S. Department of the Air Force 2001).

Gibson et al. (2001a) examined the medical records of Air Force personnel occupationally exposed to aircraft fuel and compared them with records of unexposed (control) personnel. The exposed group consisted of 5,706 people (242 women and 5,464 men), and the control group consisted of 5,706 people (2,853 women and 2,853 men) randomly selected from a cohort of 20,244 Air Force unexposed personnel. Preliminary results showed that the total numbers of medical visits and visits for specific reasons, including respiratory problems, were not markedly different among the exposed and unexposed groups. Specific diseases, including respiratory illnesses, were examined, but no marked differences were found between the groups. This study is limited by many factors, including limited information on potential confounders, completeness of health-event recording, differences among personnel in availability of health care, consequences of taking sick leave for health-care visits, differences in health-care-seeking behavior, and differences in amount of self-care or sensitivity to symptoms of illness.

Gibson et al. (2001b) conducted a health survey of 328 Air Force personnel (276 men and 52 women) with a self-assessment questionnaire. The subjects' exposures were categorized as high (performed duties associated with aircraft fuel systems), moderate (may have come into contact with jet fuel in the course of their duties), or low (did not normally come into contact with jet fuel or other solvents while performing their duties). Only one of the several measures evaluated, "difficulty breathing," was related to the effects of jet fuel on the respiratory tract. Preliminary results showed no statistical differences (adjusted for age, gender, and smoking history) between the high- and moderate-exposure groups compared with the low-exposure group in the reported symptom of "difficulty breathing." This study is limited by the fact that the symptoms were self-reported, allowing for bias.

In a study of sensory threshold of deodorized kerosene in humans, six volunteers 20 to 63 years (yr) old, were exposed to kerosene at 140 mg/m^3 (20 ppm) for 15 min. No eye, nose, or throat irritation was reported during or after exposures. Three volunteers reported slight olfactory fatigue (Carpenter et al. 1976).

Recent case reports suggest that prolonged exposure to kerosene vapors may result in development of asthma and other respiratory tract symptoms (Todd and Buick 2000). Three families (six adults, three children) were exposed to kerosene vapors for 4-8 months as a result of a spill near their homes. Exposures occurred for an estimated 100 hr/wk. Concentrations in one home were measured at 5.6-79.7 mg/m^3. Three of the children and one adult developed asthma that persisted for more than 2 yr. The remaining adults developed other respiratory tract symptoms, such as sore throat, cough, watery eyes, stuffed noses, and chest tightness. That study is limited because only a small group of people (n = 9) were exposed.

EFFECTS OF EXPOSURE TO JET FUELS AND KEROSENE IN EXPERIMENTAL ANIMALS

Several animal inhalation toxicity studies have been conducted on various jet fuels (summarized in Table 4-2). In one study, male F344 rats were exposed to shale-oil-derived JP-4 continuously for 90 days by inhalation at 1,000 mg/m^3. The exposure resulted in no effects on lung volumes, dynamic resistance and compliance, quasistatic compliance, partial and full forced vital capacities, carbon monoxide diffusion capacity, and closing volume. There

TABLE 4-2 Effects of Jet Fuel Exposure on the Respiratory Tract in Experimental Animals

Fuel Type	Species or Cell Line	Exposure Concentration	Exposure Duration	Effects	Reference
JP-4	F344 rats	1,000 mg/m^3	90 days continuously	Exposure resulted in no effects on lung volumes, dynamic resistance and compliance, quasistatic compliance, partial and full forced vital capacities, carbon monoxide diffusion capacity, and closing volume; no effects on deposition or clearance of inhaled ^{51}Cr-labeled microspheres; no evidence of pulmonary disease in control and exposed rats	Newton et al. 1991
JP-8	F344 rats, C57BL/6 mice	500, 1,000 mg/m^3 (vapor)	90 days	No respiratory tract effects attributed to JP-8; results well characterized with regard to concentration and chemical composition	Mattie et al. 1991
JP-8	F344 rats	495-520, 813-1,094 mg/m^3 (aerosol-vapor mixture)	1 hr/day, 5 days/wk for 7, 28, 56 days	Pulmonary resistance increased in 7- and 28-day exposure groups; lung-permeability data indicated lung injuries peaking at 28 d of exposure; all groups had interstitial edema resulting from endothelial damage; all groups had activated thickening of alveolar septa and alveolar macrophages	Hays et al. 1995; Pfaff et al. 1995
JP-8	C57BL/6, B6.A.D. (Ahr^d/Nat^s knockout)	Up to 118 mg/m^3 aerosol; vapor concentration	1 hr/day for 7 days	Exposures resulted in increases in total protein and LDH among groups at high concentrations; minimal morphologic changes after inhalation; damage to bronchiolar	Robledo and Witten 1998; Robledo et al. 2000; Wang et

	mice	not reported		epithelium resulting in perivascular edema and damaged Clara cells at highest concentrations	al. 2001
JP-4, JP-8, and JP-8 +100	Swiss-Webster mice	JP-4: 685-11,430 mg/m^3, JP-8: 681-3,613 mg/m^3, JP-8 + 100: 777-2,356 mg/m^3 (vapor and aerosol)	30 min	Exposure to jet fuels caused breathing patterns characteristic of upper airway sensory irritation at all concentrations but no apparent deep lung irritation at any concentration; RD_{50} determined to be 4,842 mg/m^3 for JP-4, 2,876 mg/m^3 for JP-8, 1,629 mg/m^3 for JP-8 +100	US Department of the Air Force 2001
Deodorized kerosene	Rats, beagles	20, 48, 100 mg/m^3 (vapor)	6 hr/day, 5 days/wk for up to 67 days	No pulmonary lesions observed in either species	Carpenter et al. 1976
Kerosene	Rabbits, guinea pigs	31,000-35,000 mg/m^3 (aerosol)	15 min/day for 21 days	Short-term exposures resulted in significant increases in total pulmonary resistance, increased tracheal resistance in rabbits; tracheas of exposed guinea pigs showed enhanced susceptibility to acetylcholine immediately and 24 hr after exposure; damage to tracheal ciliated cells in guinea pigs observed; increased numbers of nucleated and epithelial cells recovered in BALF	Casacó et al. 1982; Casacó et al. 1985a,b; Noa and Sanabria 1984; Noa et al. 1985

(Continued)

TABLE 4-2 *Continued*

Fuel Type	Species or Cell Line	Exposure Concentration	Exposure Duration	Effects	Reference
JP-8	In vitro cell lines: rat lung alveolar type II epithelial cells, human histocytic lymphoma cells, T-cell leukemia cells	80 µg/ml	24 hr	Rat lung cell line induced biochemical and morphologic markers of apoptotic cell death; T-cell leukemia cell lines resistant to cytotoxic effects of JP-8	Stoica et al. 2001
JP-8	Swiss-Webster mice	1,000, 2,500 mg/m^3 (vapor-aerosol mixture)	1 hr/day for 7 days	Of 796 proteins analyzed, 42 were altered by exposure to JP-8 at 2,500 mg/m^3. 8 were increased, 34 were decreased in abundance; 1 of 42 proteins altered at 2,500 mg/m^3 was also altered at 1,000 mg/m^3	Witzmann et al. 1999

Abbreviations: BALF, bronchoalveolar lavage fluid; LDH, lactate dehydrogenase; RD_{50}, dose that causes 50% decrease in respiratory rate; ^{99m}Tc-DTPA, technetium-99m diethylenetriamine pentaacetic acid.

were no effects on deposition or clearance of ^{51}Cr-labeled microspheres inhaled later. Lungs of the control and JP-4-exposed rats showed no evidence of pulmonary disease (Newton et al. 1991).

Continuous inhalation of JP-8 vapor by F344 rats and C57BL/6 mice at 500 and 1,000 mg/m^3 for 90 days resulted in no respiratory tract effects that could be attributed to JP-8 (Mattie et al. 1991). The exposures were well characterized with regard to concentration and chemical composition.

Studies were conducted on the effects of JP-8 inhalation on the rat lung (Hays et al. 1995; Pfaff et al. 1995). Male F344 rats were exposed to an aerosol and vapor mixture of JP-8 at 495-520 mg/m^3 (lower concentration) or 813-1,094 mg/m^3 (higher concentration) for 1 hr/day, 5 days/wk for 7, 28, and 56 days. The aerosol particles were respirable and tended to be monodisperse (mass median aerodynamic diameter, 1.08-1.51; geometric standard deviation, 1.5-2.2). The aerosol:vapor ratio was reported at 1.2-1.8 (mean, 1.5). Control groups were sham-exposed for 7, 28, and 56 days. The lower-exposure concentration caused an increase in dynamic compliance after 7 days, but the effect did not persist with continued exposure to 28 days (Pfaff et al. 1995). Pulmonary resistance was increased in both the 7- and 28-day fuel-exposed groups. Pulmonary function was not measured in the low-exposure group at 56 days or in the high-exposure group at any time (Hays et al. 1995). Lung epithelial permeability, as determined by pulmonary clearance of technetium-^{99m}labeled diethylenetriamine pentaacetate (^{99m}TcDTPA), was measured in all rats. Clearance rates were significantly increased above the corresponding time-control value among the low-exposure group after 7 days (Pfaff et al. 1995), the low- and high-exposure groups after 28 days, and the high-exposure group after 56 days (Hays et al. 1995). Lung-permeability data indicated that lung injuries peaked at 28 days of fuel exposure. Although no treatment-related pulmonary lesions observable by light microscopy were reported in rats exposed to JP-8 at 500 mg/m^3 for up to 28 days (Pfaff et al. 1995), electron micrographs showed that all groups had interstitial edema resulting from endothelial damage. There was an apparent thickening of the alveolar septa, and alveolar macrophages were activated in all groups (Hays et al. 1995). Lung-permeability data correlated with histology data.

In followup experiments, the pulmonary effects of JP-8 inhalation in mice were evaluated (Robledo and Witten 1998; Robledo et al. 2000; Wang et al. 2001). Groups of C57BL/6 and B6.A.D. (Ahr^d/Nat^s knockout) mice were exposed 1 hr/day for 7 days to ambient air or aerosolized JP-8 at 0-118 mg/m^3. JP-8 vapor was also present; however, the concentration of the vapor in the exposure atmosphere was not reported. Exposure had no effect on dynamic compliance or resistance. Lung epithelial permeability, as determined with ^{99m}TcDTPA, was affected in C57BL/6 mice exposed at 50 and 113

mg/m^3 and in B6.A.D. (Ahr^d/Nat^s) mice exposed at 48 (but not 50), 113, and 118 mg/m^3. Exposures resulted in increases in total protein, an indicator of pulmonary edema, and lactate dehydrogenase (LDH), an indicator of pulmonary cell damage or death, among groups exposed at the higher JP-8 concentrations, but results varied between mouse strains and experiments. Conflicting effects of exposure on *N*-acetyl-b-D-glucosaminidase and on numbers of nucleated cell numbers and profiles recovered with bronchoalveolar lavage were reported for the two sets of experiments. For example, significant increases in nucleated cell counts were reported for C57Bl/6 and B6A.D. (Ahr^dNar^s) mice exposed to JP-8 at 113 mg/m^3 (Robledo and Witten 1998), but significant decreases were reported for B6A.D. (Ahr^dNar^s) mice exposed to JP-8 at 48 and 118 mg/m^3 (Robledo et al. 2000). Minimal morphologic changes observed with light microscopy were reported after JP-8 inhalation, but ultrastructural evaluations revealed damage to bronchiolar epithelium resulting in perivascular edema and damage to Clara cells in mice exposed at the highest concentrations.

Early work on the effects of JP-8 inhalation in rats showed an inverse relationship between increases in airway epithelial permeability (^{99}TcDTPA clearance) and decreased concentrations of the tachykinin substance P in bronchoalveolar lavage fluid (BALF) (Hays et al. 1995). Substance P has a strong affinity for the neurokinin receptor NK_1, one of a family of plasma-membrane-bound neurokinin receptors that mediate protective reflex responses—such as bronchoconstriction, increased vascular permeability, vasodilatation, mucus secretion, and enhanced mucociliary activity—to airway exposure to mechanical or chemical irritants.

Robledo and Witten (1999) further examined the role of substance P and receptor NK_1 in mediating JP-8-induced lung injury. Groups of Ahr^d/Nat^s mice were exposed to JP-8 at 50 mg/m^3 for 1 hr/day for 7 days. End points included of pulmonary function, biochemical and cellular changes in BALF, ^{99m}TcDPTA lung-clearance rates, and pulmonary morphology (based on light and electron microscopic evaluations). The effects of administration of an aerosol of (Sar^9,$Met[O_2]^{11}$) substance P daily immediately after JP-8 inhalation and the effects of daily injection of the NK_1 receptor antagonist CP-96345 on the end points were determined in additional groups of mice. JP-8 inhalation had no effect on pulmonary dynamic compliance or resistance. Exposure significantly increased ^{99m}Tc-DTPA clearance rate and total protein and LDH in lavage fluid. Exposure significantly decreased *N*-acetyl-b-D-glucosamine and total numbers of nucleated cells and macrophages in BALF. Morphologically, JP-8 induced marked focal areas of alveolar septal thickening and collapsed air spaces. Distal airways were characterized by the appearance of swollen and exfoliated bronchiolar epithelial cells. Areas of vacuolization between bronchioles and venules suggestive of subendothelial edema were also observed. Mice given substance P by inhalation showed none of the adverse functional, biochemical, or morphologic changes induced by JP-8 inhalation. However, the JP-8-induced adverse effects were exacerbated by administration of the NK_1 antagonist CP-96345. Results of the studies suggest that JP-8 inhalation triggers a protective response in the lung, mediated by a substance P/NK_1 receptor that may become overwhelmed by repeated inhalation of the fuel at high concentrations.

The results of the pulmonary effects of inhaled JP-8 in mice and rats should be viewed with caution because exposure atmospheres are not well defined with respect to the concentration of vapor coexisting with aerosols. Morphologic evaluations were conducted in only a few animals per exposure group and were not reported quantitatively with regard to incidence or semiquantitatively with respect to severity. Pulmonary function measurements were not performed with conventional methods. Furthermore, conflicting results were obtained in replicated experiments with mice.

The sensory-irritation potential of JP-4, JP-8, and JP-8+100 were evaluated in groups of four male Swiss-Webster mice exposed, head-only, for 30 min to atmospheres of each material containing both vapor and aerosol phases (U.S. Department of the Air Force 2001). The three test materials evoked breathing patterns characteristic of upper airway sensory irritation at all exposure concentrations. Examination of the breathing patterns revealed no apparent pulmonary (deep lung) irritation at any concentration. The calculated values for concentrations of JP-4, JP-8, and JP-8+100 that caused a 50% decrease in respiratory rate (RD_{50}) were 4,842, 2,876, and 1,629 mg/m^3 (total aerosol + vapor concentration), respectively. The relative irritancy ranking of the three fuels was JP-8+100 > JP-8 > JP-4. In contrast, respiratory rates of mice exposed to deodorized kerosene vapor by inhalation at 6,900 mg/m^3 were not decreased by 50% or more from control values (Carpenter et al. 1976).

To determine the effects of repeated inhalation of deodorized kerosene vapor, groups of 25 male rats and four male beagles were exposed at 20, 48, or 100 mg/m^3 for 6 hr/day, 5 days/wk for up to 67 days. Characteristics examined included histopathologic findings, serum chemistry, and electrocardiograms (dogs only). No pulmonary lesions were observed in either species. No histopathologic evaluation of the upper respiratory tract was performed (Carpenter et al. 1976).

Studies on the effects of short-term inhalation of kerosene aerosol (about 8 mm in diameter) at 3,100-3,500 mg/m^3 have been conducted. Short-term exposures resulted in significance increases in total pulmonary resistance and increased tracheal resistance in rabbits (Casacó et al. 1982). Tracheas from exposed guinea pigs showed enhanced susceptibility to acetylcholine immediately and 24 hr after exposure (Casacó et al. 1985b). Repeated inhalation by guinea pigs (15 min/day for 21 days) resulted in damage to tracheal ciliated cells (Noa and Sanabria 1984) and pulmonary inflammatory responses and toxicity, as indicated by increases in numbers of nucleated and epithelial cells recovered in BALF (Noa et al. 1985).

The effects of JP-8 inhalation on protein expression in mouse lung were evaluated in Swiss-Webster mice (Witzmann et al. 1999). Mice were exposed to an aerosol and vapor mixture of JP-8 at 1,000 or 2,500 mg/m^3 for 1 hr/day for 7 days. Lung cytosol was prepared and analyzed. An average of 796 proteins were resolved in each sample pattern and matched reproducibly to a reference pattern. Of the 796 proteins, 42 were significantly altered ($p < 0.001$) by exposure to JP-8 at 2,500 mg/m^3. Eight proteins were increased and 34 decreased in abundance. One of the 42 proteins altered at 2,500 mg/m^3 was altered at 1,000 mg/m^3, but 11 were altered significantly ($p < 0.01$). The observed alterations suggested four general effects: impaired synthetic and processing machinery, ultrastructural damage, toxic and metabolic stress and detoxification systems, and functional responses to CO_2 handling, acid-base homeostasis, and fluid secretion. The effects were considered to be consistent with morphologic and functional alterations observed in mice after JP-8 exposure. The relevance of the findings to evaluating the scientific basis of the interim permissible exposure level (PEL) of 350 mg/m^3 is questionable considering the high JP-8 concentration used in the study and the fact that the vapor concentration was not characterized.

EFFECTS OF IN VITRO EXPOSURE TO JP-8

Stoica et al. (2001) investigated apoptosis as the molecular mechanism responsible for the cellular toxicity induced by JP-8 in several cell lines. JP-8 exposure of a rat lung alveolar type II epithelial cell line (RLE-6TN) induced biochemical and morphologic markers of apoptotic cell death, including caspase-3 activation, poly (ADP-ribose) polymerase cleavage, chromatin condensation, membrane blebbing, cytochrome c release from mitochondria, and genomic DNA cleavage into both oligonucleosomal (DNA ladder) and high-molecular-weight fragments. Similar responses to JP-8 were seen in the human histocytic lymphoma cell line (U937) and the T-cell leukemia cell line (Jurkat). Jurkat cells stably transfected with a plasmid encoding the antiapoptotic protein Bcl-x_L or pretreated with the pan-caspase inhibitor Boc-D-frnk were resistant to the cytotoxic effects of JP-8. The results suggested that apoptotic cell death was at least partially responsible for the cytotoxic effects of JP-8. The apoptotic effects were seen with JP-8 dilutions of 1×10^{-4}. Assuming a JP-8 density of 1 g/mL, that translates into concentrations of approximately 100 mg/mL of cell incubation medium. The physiologic relevance of that high concentration is questionable, so it is difficult to determine the relevance of the results to human exposure.

CONCLUSIONS AND RECOMMENDATIONS

Preliminary comparison of medical records of Air Force personnel occupationally exposed to JP-8 with records of unexposed (control) personnel showed that numbers of medical visits related to respiratory problems were not markedly different between the exposed and unexposed groups. Specific diseases, including respiratory illnesses, were examined, but no marked differences were found between the groups.

No respiratory-tract effects were found in F344 rats and C57BL/6 mice exposed to JP-8 vapor at 500 or 1,000 mg/m^3 for 90 days. However, several animal studies conducted in F344 rats and C57BL/6 mice suggest that mixtures of JP-8 vapors and aerosols can result in pulmonary inflammation and alterations in pulmonary functions. Toxic effects have been reported in C57BL/6 mice exposed at concentrations as low as 50 mg/m^3 for 1 hr per day for 7 days. The results from those studies suggest that JP-8 aerosol is more toxic to the respiratory tract than JP-8 vapor. The subcommittee reviewed the methods used to generate the exposure atmospheres in the studies using mixtures of vapors and aerosols and suspects that the JP-8 concentrations in the atmosphere may have been underreported. However, even if the actual concentration was 20 times as high (i.e., if exposure was at a concentration of 1,000 mg/m^3), the observation of positive effects from a short exposure duration (1 hr/day for 7 days) at that concentration leads the subcommittee to conclude that the interim PEL of 350 mg/m^3 might be too high to be protective of human health (assuming the application of commonly used uncertainty factors).

Because there are concerns about the characterization of the exposure atmospheres in the studies using mixtures of vapors and aerosols, the subcommittee recommends an examination of the methods of characterizing the exposure atmosphere. Future studies involving exposures to aerosols should be designed in collaboration with scientists knowledgeable in aerosol generation, aerosol physics, and quantification of vapors and aerosols to ensure accurate characterization of exposure atmospheres. The subcommittee recommends that respiratory-system toxicity be evaluated in experimental animals exposed to JP-8 vapors and mixtures of vapors and aerosols by the inhalation route. Because the composition of JP-8 varies from batch to batch, scientists with expertise in petroleum toxicology should be consulted to design the best approach for testing the respiratory-system toxicity of JP-8 (e.g., testing JP-8 samples at the extremes of their composition ranges or testing JP-8 samples so that the concentrations of component classes can be correlated with toxic end points).

REFERENCES

Carpenter, C.P., D.L. Geary Jr., R.C. Myers, D.J. Nachreiner, L.J. Sullivan, and J.M. King. 1976. Petroleum hydrocarbon toxicity studies. XI. Animal and human response to vapors of deodorized kerosene. Toxicol. Appl. Pharmacol. 36(3):443-456.

Casacó, A., D. Carvajal, M. Noa, R. González, M. García, and A.R. de la Vega. 1985a. Effects of kerosene on airway sensitization to egg albumin in guinea pig. Allergol. Immunopathol. 13(3):235-239.

Casacó, A., R. González, L. Arruzazabala, M. García, and A.R. de la Vega. 1985b. Kerosene aerosol induces guinea-peg airway hyperreactivity to acetylcholine. Respiration 47(3):190-195.

Casacó, A., R. González, L. Arruzazabala, M. García, and A.R. de la Vega. 1982. Studies on the effects of kerosine aerosol on airways of rabbits. Allergol. Immunopathol. 10(5):361-366.

Gibson, R.L., S. Shanklin, and R.L. Warner. 2001a. Health effects comparisons. Pp. 125-129 in JP-8 Final Risk Assessment Report. The Institute of Environmental and Human Health (TIEHH), Lubbock, TX. August 2001.

Gibson, R.L., S. Shanklin, and R.L. Warner. 2001b. Self-reported health status. Pp. 132-139 in JP-8 Final Risk Assessment Report. The Institute of Environmental and Human Health (TIEHH), Lubbock, TX. August 2001.

Hays, A.M., G. Parliman, J.K. Pfaff, R.C. Lantz, J. Tinajero, B. Tollinger, J.N. Hall, and M.L. Witten. 1995. Changes in lung permeability correlate with lung histology in a chronic exposure model. Toxicol. Ind. Health 11(3):325-336.

Knave, B., B.A. Olson, S. Elofsson, F. Gamberale, A. Isaksson, P. Mindus, H.E. Persson, G. Struwe, A. Wennberg, and P. Westerholm. 1978. Long-term exposure to jet fuel. II. A cross-sectional epidemiologic investigation on occupationally exposed industrial workers with special reference to the nervous system. Scand. J. Work Environ. Health 4(1):19-45.

Mattie, D.R., C.L. Alden, T.K. Newell, C.L. Gaworski, and C.D. Flemming. 1991. A 90-day continuous vapor inhalation toxicity study of JP-8 jet fuel followed by 20 or 21 months of recovery in Fischer 344 rats and C57BL/6 mice. Toxicol. Pathol. 19(2):77-87.

Newton, P.E., S.V. Becker, and C.J. Hixon. 1991. Pulmonary function and particle deposition and clearance in rats after a 90-day exposure to shale-oil-derived jet fuel JP-4. Inhal. Toxicol. 3(2):195-210.

Noa, M., and J. Sanabria. 1984. Tracheal ultrastructure in kerosine treated guinea pigs: A preliminary report. Allergol. Immunopathol. 12(1):33-36.

Noa, M., J. Illnait, and R. González. 1985. Cytologic and biochemical changes in pulmonary washings of guinea pigs exposed to kerosene. Allergol. Immunopathol. 13(3):193-196.

NRC (National Research Council). 1996. Permissible Exposure Levels for Selected Military Fuel Vapors. Washington, DC: National Academy Press.

Pfaff, J., K. Parton, R.C. Lantz, H. Chen, A.M. Hays, and M.L. Witten. 1995. Inhalation exposure to JP-8 jet fuel alters pulmonary function and substance P levels in Fischer 344 rats. J. Appl. Toxicol. 15(4):249-256.

Robledo, R.F., and M.L. Witten. 1998. Acute pulmonary response to inhaled JP-8 jet fuel aerosol in mice. Inhal. Toxicol. 10(5):531-553.

Robledo, R.F., and M.L. Witten. 1999. NK1-receptor activation prevents hydrocarbon-induced lung injury in mice. Am. J. Physiol. 276(2 Pt 1):L229-L238.

Robledo, R.F., R.S. Young, R.C. Lantz, and M.L. Witten. 2000. Short-term pulmonary response to inhaled JP-8 jet fuel aerosol in mice. Toxicol. Pathol. 28(5):656-663.

Stoica, B.A., A.H. Boulares, D.S. Rosenthal, S. Iyer, I.D. Hamilton, and M.E. Smulson. 2001. Mechanisms of JP-8 jet fuel toxicity. 1. Induction of apoptosis in rat lung epithelial cells. Toxicol. Appl. Pharmacol. 171(2):94-106.

Todd, G.R.G., and B. Buick. 2000. Asthma due to kerosine exposure: Three case reports. Int. J. Occup. Med. Environ. Health 13(1):23-25.

Tunnicliffe, W.S., S.P. O'Hickey, T.J. Fletcher, J.F. Miles, P.S. Burges, and J.G. Ayres. 1999. Pulmonary function and respiratory symptoms in a population of airport workers. Occup. Environ. Med. 56(2):118-123.

U.S. Department of the Air Force. 2001. Sensory Irritation Study in Mice. Final Report. Project Number 162951. Test Substance: JP-4(MRD-00-629), JP-8(MRD-00-630), JP-8+100 (MRD-00-631). Prepared by ExxonMobil Biomedical Sciences, Inc., Annandale, NJ, for the U.S. Department of the Air Force, Brooks Air Force Base, TX.

Wang, S.J., R.S. Young, and R.L. Witten. 2001. Age-related differences in pulmonary inflammatory responses to JP-8 jet fuel aerosol inhalation. Toxicol. Ind. Health 17(1):23-29.

Witzmann, F.A., M.D. Bauer, A.M. Fieno, R.A. Grant, T.W. Keough, S.E. Kornguth, M.P. Lacey, F.L. Siegel, Y. Sun, L.S. Wright, R.S. Young, and M.L. Witten. 1999. Proteomic analysis of simulated occupational jet fuel exposure in the lung. Electrophoresis 20(18):3659-3669.

5

Effects of Jet-Propulsion Fuel 8 on the Nervous System

This chapter summarizes the findings on potential neurotoxicity from exposure to jet-propulsion fuel 8 (JP-8) presented in the National Research Council report *Permissible Exposure Levels for Selected Military Fuel Vapors* (NRC 1996) and reviews additional studies, most of which were completed after the 1996 report was published. Since the 1996 report was released, additional epidemiologic studies associated with occupational JP-8 exposure and experimental animal studies examining the neurotoxic potential of kerosene-based jet fuels, including JP-8, and kerosene via the dermal and inhalation routes have been conducted. The subcommittee used the available information on JP-8 to assess the potential for toxic effects of JP-8 on the nervous system in humans.

SUMMARY OF STUDIES DISCUSSED IN THE 1996 NATIONAL RESEARCH COUNCIL REPORT

The National Research Council Subcommittee on Permissible Exposure Levels for Military Fuels reviewed studies on the potential toxic effects of JP-5, JP-8, and diesel fuel marine (DFM) on the nervous system (NRC 1996). The vapors from those fuels contain a mixture of volatile hydrocarbons, which at high concentrations are central nervous system (CNS) depressants and can produce anesthesia or asphyxia at high absorbed doses (Andrews and Snyder 1986; Marshall and Wollman 1985). The effectiveness of vapors as CNS depressants depends principally upon the volatility of their component hydrocarbons.

The Subcommittee on Permissible Exposure Levels for Military Fuels found that data on potential nervous system effects of jet fuels are sparse. In several Swedish studies conducted by Knave and his colleagues, acute CNS symptoms were reported in workers who were employed in jet factories where they were potentially exposed to jet fuels designated Jet A-1 and JP-1 (Knave et al. 1976, 1978, 1979). Industrial-hygiene measurements of up to 3,200 mg/m^3 were reported for a variety of job activities. Although the one-time air measurements reflected various activities, the exposures were not well characterized over time or by individual.

In a study of 30 Swedish workers potentially exposed to jet fuels at a motor factory for an average of 17 years (yr), workers reported acute symptoms of exposure to vapors and performance degradation associated with long-term exposure (Knave et al. 1978). The study reported an approximate time-weighted average (TWA) of 300 mg/m^3. The findings of performance degradation said to be attributable to long-term exposure were considered unreliable for a number of reasons, including weak and inconsistent evidence of impairment, inadequate methods of evaluation, inadequate consideration of confounding, a small cohort of workers, and a lack of quantitative information on exposure over time.

EFFECTS OF EXPOSURE TO JP-8 IN HUMANS

Acute exposure to jet fuels has been associated with neurologic effects in humans, including headache, nausea, vomiting, dizziness, fatigue, in coordination, irritability, problems with attention and memory, narcosis, and gait disturbances (Knave et al. 1976; Knave et al. 1978; Porter 1990; Anger and Storzbach 2001; Gibson et al. 2001b) (see Table 5-1). Persistent effects can include peripheral neuropathy and behavioral changes, such as reduced performance on tests of attention and psychomotor speed.

In a preliminary assessment of data, Anger and Storzbach (2001) reported significant behavioral disturbances characterized by impaired performance on digit-span (forward), digit-symbol, and finger-tapping tests among workers who had high JP-8 exposure at the beginning of their workshifts compared with workers who had no significant JP-8 exposure. Exposure was determined by median breathing-zone concentrations of two components of JP-8,

TABLE 5-1 Effects of JP-8 on the Nervous System in Humans

Reference	Exposure Concentration	Exposure Duration	Study Results
Knave et al. 1976; 1978; 1979[a]	Overall average concentration, 300 mg/m^3 (range, 85-974 mg/m^3)	Average employment duration of 17 yr	21 of 30 workers reported recurrent acute symptoms on exposure; exposed workers reported higher prevalence of neurasthenic symptoms, greater irregularity of performance on test of complex reaction time, greater performance decrement over time in simple reaction-time task, poorer performance in task of perceptual speed than control group
Anger and Storzbach 2001[b]	Measurements taken in breathing zones of subjects; median concentration of naphthalene, 1.9 μg/m^3 (low-exposure group), 447 μg/m^3 (high-exposure group); median concentration of benzene, 3.1 μg/m^3 (low-exposure group), 242 μg/m^3 (high-exposure group)	High-exposure group had persistent exposure to JP-8 (defined as 1 hr twice per wk for at least 9 mo); low-exposure group had no significant exposure to jet fuel or solvents	Subjects were given seven neurobehavioral tests in Behavioral Assessment and Research System; before exposure, high-exposure group had significantly lower performance on digit-span forward and backward test, symbol digit-latency test, and tapping test than low-exposure group; results of tests did not correlate with breath or passive naphthalene or benzene exposure; effects may be result of carryover from previous exposure; when pre- and post-exposure test results were compared, passive naphthalene exposure was significantly associated with performance on Oregon Dual Task Procedure, Match to Sample, and Tapping Trial 2

Bekkedal et al. 2001[b]	Measurements taken in breathing zones of subjects; median concentration of naphthalene, 1.9 $\mu g/m^3$ (low-exposure group), 447 $\mu g/m^3$ (high-exposure group); median concentration of benzene, 3.1 $\mu g/m^3$ (low-exposure group) and 242 $\mu g/m^3$ (high-exposure group)	High-exposure group had persistent exposure to JP-8; low-exposure group had no significant exposure to jet fuel or solvents	Subjects were given eye-blink conditioning response test; morning session showed that high-exposure group had statistically significant differences in percentage CR, CR peak latency, and CR onset latency; high-exposure group also had fewer CRs than low-exposure group; no statistically significant exposure-based differences afternoon session showed
Bhattacharya et al. 2001[b]	Measurements taken in breathing zones of subjects; median concentration of naphthalene, 1.9 $\mu g/m^3$ (low-exposure group), 10.4 $\mu g/m^3$ (moderate-exposure group), 447 $\mu g/m^3$ (high-exposure group); median concentration of benzene, 3.1 $\mu g/m^3$ (low-exposure group), 7.45 $\mu g/m^3$ (moderate-exposure group), 242 $\mu g/m^3$ (high-exposure group)	High- and moderate-exposure groups had persistent exposure to JP-8; low-exposure group had no significant exposure to jet fuel or solvents	Subjects were given postural sway tests; post-log sway length, based on ANCOVA analysis after controlling for cofactors, was significantly associated with passive naphthalene exposure for "eyes closed no foam" and "eyes closed bending" tests
Gibson et al. 2001a[b]	Exposed group (5,706) had potential occupational exposure to JP-8; control group (5,706) did not work in occupations in which exposure to JP-8 would occur; all subjects were active duty members of U.S. Air Force	Not reported	Review of medical records showed no differences between exposed and control groups in neurologic and mental illnesses

(Continued)

TABLE 5-1 *Continued*

Reference	Exposure Concentration	Exposure Duration	Study Results
Gibson et al. 2001b[b]	Measurements taken in breathing zones of subjects; median concentration of naphthalene, 1.9 μg/m^3 (low-exposure group), 10.4 μg/m^3 (moderate-exposure group), 447 μg/m^3 (high-exposure group); median concentration of benzene, 3.1 μg/m^3 (low-exposure group), 7.45 μg/m^3 (moderate-exposure group), 242 μg/m^3 (high-exposure group)	High- and moderate-exposure groups had persistent exposure to JP-8; low-exposure group had no significant exposure to jet fuel or solvents	In self-assessment questionnaire, subjects in high- and moderate-exposure groups reported more headaches, dizziness, trouble concentrating, balance problems, walking difficulties, forgetfulness, and trouble in gripping objects

[a]Jet factory workers were exposed to Jet A-1 and JP-1, Swedish military's equivalent of JP-4.

[b]Data were collected from volunteers (male and female active-duty Air Force personnel) at six Air Force bases in United States. Volunteers were divided into three exposure groups: high, moderate, and low. High-exposure group performed tasks associated with repairing aircraft fuel systems; moderate-exposure group performed tasks associated with fuel handling, distribution, recovery, and testing; and low-exposure group did not routinely come into contact with jet fuel or solvents. Data were collected in morning before subjects went to work and after they completed their work for that day. Reported results are from a preliminary analysis of data. Additional background information can be found in Appendix B.

Abbreviations: ANCOVA, analysis of covariance; CR, conditioned response.

naphthalene and benzene (see Table 5-1). The low-exposure group had no specific source of exposure to JP-8, but the high-exposure group had been exposed to JP-8 occupationally for at least 9 months. The findings indicate that attention and executive function, working memory, and psychomotor function may be affected by exposure to JP-8 and that the acute effects of JP-8 on cognitive function persist in people who have relatively high exposure. The association between exposure to jet fuels and the incidence of peripheral neuropathy has been identified in reports by Knave et al. (1976, 1978). That particular finding is consistent with the proposed mechanism of action of 2,5-hexanedione and its derivatives and supports the hypothesis that exposure to jet-fuel constituents may affect nervous system functioning because of the formation of a metabolite that can react with cellular macromolecules to induce neuropathy (Anthony et al. 1983a; Anthony et al. 1983b; Graham et al. 1995).

Preliminary data analyses show disturbances of balance among subjects exposed to jet fuels that may reflect reversible depression of CNS function and disturbances of peripheral sensory perception due to neuropathy and disruption of cerebellar function (see Table 5-1 and the section in Appendix C on posturograms) (Bhattacharya 2001). A preliminary analysis of data by Bekkedal et al. (2001) suggests that the eyeblink conditioning response may be affected by exposure to JP-8 (see Table 5-1 and the section in Appendix C on blink-reflex classical conditioning). Several constituents of JP-8—toluene and xylene—are known to have neurotoxicologic effects in humans. It is not known whether exposure to such chemicals at the concentrations found in JP-8 will cause adverse neurologic effects and whether their presence in the mixture produces additive, synergistic, or antagonistic effects.

EFFECTS OF EXPOSURE TO JP-8 IN EXPERIMENTAL ANIMALS

This section describes experimental-animal studies that have assessed the neurotoxic potential of JP-8 and related fuels. The studies are summarized in Table 5-2.

Baldwin et al. (2001) exposed 6-month-old Fischer 344 rats to room air or JP-8 aerosols alone or to JP-8 and then aerosolized substance P, which has been shown to attenuate the effects of JP-8-induced pulmonary dysfunction and immunotoxicity in animals. Inhalation exposures were nose-only and performed 1 hr/day, 5 days/wk for 28 days. Aerosolized JP-8 with a mass mean aerodynamic diameter (MMAD) of 1.7-1.9 μm (M. Witten, University of Arizona, personal communication, 2002) was administered to the rats at

TABLE 5-2 Effects of Jet Fuel on Nervous System in Experimental Animals

Fuel Type	Species	Exposure Concentration	Exposure Duration	Effects	Reference
JP-8 (aerosol)	F344 rats	1,059 mg/m^3 for first 25 days, 2,491 mg/m^3 for final 3 days	1 hr/day, 5 days/wk for 28 days	Neurologic measures were assessed with functional observation battery; exposed rats had significant differences in spontaneous activity and CNS excitability from controls; exposed rats exhibited greater velocity of swimming in Morris swimming task	Baldwin et al. 2001
JP-8 (vapor)	Sprague-Dawley rats	1,000 or 5,000 mg/m^3	6 hr/day, 5 days/wk for 6 wk, followed by no exposure for 64 days	High-dose group was significantly impaired relative to low-dose group in difficult task involving pressing one or more levers after auditory cue and in task involving complicated repeated acquisition; no differences observed between two groups in simple autoshaping and fixed-ratio or spatial-reversal tasks; low-dose group exhibited superior performance relative to control group in test requiring three or four lever presses in three-lever array	Ritchie et al. 2000, 2001a,b
JP-8, JP-5 (vapor)	Sprague-Dawley rats	1,000 (JP-8) or 1,200 (JP-5) mg/m^3	6 hr/day, 5 days/wk for 6 wk, followed by no exposure for 65 days	Significant differences were observed in JP-8-exposed group in appetitive reinforcer approach sensitization compared with JP-5-exposed group and control group; JP-5-exposed group showed increased forelimb grip strengths compared with JP-8-exposed	Rossi et al. 2001

				group and the control group; neurotransmitter concentrations were also measured; JP-8 exposure was associated with decreased concentrations of 3,4-dihydroxyphenylacetic in cerebellum and brainstem; JP-5 exposure was associated with increased concentrations of dopamine and 3,4-dihydroxyphenyl-acetic acid in hippocampus and cortex, respectively, and with decreased concentrations of homovanillic acid in hippocampus; blood samples contained increased and decreased concentrations of 5-hydroxyindoleacetic acid in JP-5 and JP-8 exposed groups, respectively	
JP-4 (vapor)	Sprague-Dawley rats	2,000 mg/m^3	6 hr/day for 14 days, followed by no exposure for 14 days or 60 days	Significant increase in appetitive reinforcer approach sensitization was observed for in short-recovery-period group, but not long-recovery-period group; long-recovery-period group exhibited significant differences in prepulse inhibition trial and treadmill response compared with controls and decrease in total locomotor activity compared with short-recovery-period group; no other differences in neurologic measures were observed; blood serotonin concentrations were increased in short-recovery-period	Nordholm et al. 1999

(Continued)

TABLE 5-2 *Continued*

Fuel Type	Species	Exposure Concentration	Exposure Duration	Effects	Reference
				group; blood 5-hydroxyindoleacetic acid was significantly increased in short- and long-recovery-period groups; serotonin and 5-hydroxyindoleacetic acid concentrations were increased in short- and long-recovery-period groups in cerebellum, brainstem, and hippocampal regions; those chemicals were also increased in striated region in short-recovery-period group and in cortical regions in long-recovery-period group	
Hydro-desulfurized kerosene	Rat	165, 330, 495 mg/kg (dermal)	5 days/wk for 13 wk	Animals were evaluated immediately after exposure period ended and after 4-wk recovery period; no significant differences were observed in functional observed battery and motor activity, startle response, and histologic evaluations	Koschier 1999

Abbreviations: CNS, central nervous system

1,059 mg/m^3 for the first 25 days and 2,491 mg/m^3 for the final 3 days. Substance P was administered at 1 μM concentration in normal saline with a nebulizer for 15 min immediately after JP-8 exposure. Neurobehavioral measures were based on functional observation battery (FOB) composed of caged and open-field observations to assess sensory, autonomic, and neuromuscular function. Spatial and visual discrimination and memory were evaluated with variations of the Morris swim task. No significant differences between the two JP-8 exposure groups were observed except in body weight. The JP-8-alone group displayed mild but significant weight loss early in the exposures but returned to pre-exposure weights by the last exposure. Because of the absence of differences in neurobehavioral measures between the JP-8 exposure groups, they were considered together and compared with controls. The JP-8-exposed rats exhibited more rearing (17 versus seven rears) for one of the five assessments performed and a greater arousal score (4.5 and 4.7 in exposed rats versus 3.8 in controls on a scale ranging from 1 to 6 with 4 considered normal) for two of the five assessments performed. Differences in the swim-task measurement were limited to greater swimming velocity in the exposed groups.

Rossi et al. (2001) exposed Sprague-Dawley rats to filtered air, JP-5 vapor at 1,200 mg/m^3, or JP-8 vapor at 1,000 mg/m^3 for 6 hr/day, 5 days/wk for 6 wks and then assessed neurobehavioral measures after a 65-day period during which there were no exposures. The neurologic tests included the acoustic-startle response, prepulse inhibition of the acoustic-startle response, appetitive-reinforcer approach sensitization, forelimb grip strength, locomotor activity, tail-flick response, conspecific approach, passive avoidance, Porsalt forced-swim test, and Morris water maze. After the neurobehavioral testing, concentrations of norepinephrine, dopamine, 3,4-dihydroxyphenylacetic acid, homovanillic acid, serotonin, and 5-hydroxyindoleacetic acid in several regions of the brain and in blood were analyzed. Significant differences were observed between the JP-8 group and the JP-5 exposure group and controls in appetitive-reinforcer approach sensitization, and JP-5-exposed animals displayed greater forelimb grip strengths than the other two groups. Significant differences in neurotransmitter concentrations were recorded relative to controls, although the perturbations in neurotransmitter concentrations were not identical in the JP-5 and JP-8 groups. The JP-5-exposed rats had increased dopamine and 3,4-dihydroxyphenylacetic acid concentrations in the hippocampus and cortex, respectively. In addition, JP-5 exposure was associated with lower concentrations of homovanillic acid in the hippocampus. In comparison, JP-8 exposure was associated with decreased 3,4-dihydroxyphenylacetic concentrations in the cerebellum and brainstem. Blood 5-hydroxyindoleacetic acid was increased and decreased in the JP-5 and JP-8 groups, respectively.

Nordholm et al. (1999) performed a study similar to that of Rossi et al. (2001). They examined the effects of repeated exposure to JP-4 on several neurobehavioral measures. Rats received whole-body exposure to JP-4 vapor at 2,000 mg/m^3 for 6 hr/day for 14 days and were then tested after a short recovery period (14 days) or a long recovery period (60 days). Neurobehavioral measures assessed were forelimb grip strength, photosensitivity, appetitive-reinforcer approach sensitization, total locomotor activity, acoustic startle and prepulse inhibition, tail-flick response, and treadmill physical fatigue. Routine histologic tests were performed on the major organs, and the same set of neurotransmitters were examined as in the Rossi et al. (2001) study. A significant increase in the appetitive-reinforcer approach sensitization was observed in the short-recovery group but not the long-recovery group relative to controls. Total locomotor activity was decreased in the long-recovery group, but not the short-recovery group relative to controls. Similarly, only the long-recovery group displayed significant differences from controls in prepulse inhibition and treadmill response relative. No other significant differences in neurobehavioral assessments were reported. Blood serotonin was higher in the short-recovery group and blood 5-hydroxyindoleacetic acid significantly higher in the short- and long-recovery group than in controls. Serotonin and 5-hydroxyindoleacetic acid were higher in the cerebellum, brainstem, and hippocampal regions in the short-recovery group and long-recovery group. Serotonin and 5-hydoxyindoleacitic acid were higher in the striated and cortical regions in the short-recovery group and the long-recovery group, respectively, than in controls.

Ritchie et al. (2000, 2001a,b) studied neurobehavioral effects of JP-8 vapor at 1,000 and 500 mg/m^3 for 6 hr/day, 5 days/wk for 6 wk followed by no exposure for 64 days. No differences were observed in simple autoshaping, fixed-ratio, or spatial-reversal tasks between exposure groups and controls. On two of 15 assessments, the high-dose group was significantly impaired relative to the low-dose group regarding a difficult task that required one or more lever presses after an auditory cue. Similarly, in the more complicated incremental repeated-acquisition task, the high-dose group exhibited significant impairment relative to the low-dose group in two of six assessments. In contrast, the low-dose group demonstrated superior performance relative to controls in a test requiring three to four lever presses in a three-lever array. This investigation suggests decreased performance in operant tasks at the highest exposure, but the significance of the findings is questionable for several reasons. There are relatively few significantly different outcomes; and, when observed, these differences occur only between the high-dose and low-dose groups and not between low-dose and control groups or high-dose and control groups. Significant differences also are observed only for one or two evaluations in a series of evaluations that otherwise demonstrate no significant differences. No dose-response relationships were demonstrated for either the neurobehavioral or the neurotransmitter measurements, and a conclusion of hormesis for the superior performance observed in the low-dose group appears premature given that only two exposure concentrations were examined.

Koschier (1999) reviewed the potential of dermal exposure to kerosene to cause adverse health effects. The author described a study in which rats were exposed to hydrodesulfurized kerosene dermally at 0, 165, 330, and 495 mg/kg for 5 days/wk for 13 wk. The rats were assessed with a FOB, and motor activity, startle response, and histologic characteristics were measured. All groups were examined after the 13-wk exposure, and the control and high-dose groups were also examined after a 4-wk recovery period. No significant differences were observed in any of the measures in any of the exposure groups.

CONCLUSIONS AND RECOMMENDATIONS

To evaluate the potential for JP-8 to cause adverse neurologic effects, the subcommittee reviewed the available data on the neurotoxicity of JP-8, related jet fuels, and kerosene in humans and experimental animals. The database on potential neurotoxicity of jet fuels is sparse, especially with regard to human studies. In an epidemiologic investigation, workers exposed to jet fuels at a Swedish jet-motor factory for an average of 17 yr were studied for possible adverse health effects. The overall TWA exposure concentration in one-time measurements was 300 mg/m^3; peak exposures were about 1,200-3,200 mg/m^3. Significant differences between exposed and unexposed workers were found with respect to nervous system effects. Most of the exposed workers reported acute symptoms, such as dizziness, headache, nausea, and fatigue. Chronic symptoms included a greater incidence of neurasthenic symptoms, such as depressed mood, lack of initiative, sleep disturbances, memory impairment, headache, dizziness, and fatigue. However, the findings of nervous system effects attributable to long-term exposure were considered questionable for a number of reasons, including weak and inconsistent evidence of impairment, inadequate methods of evaluation, inadequate consideration of confounding factors, a small cohort of workers, and a lack of quantitative information on exposure.

Preliminary results of a recent epidemiologic study on Air Force personnel occupationally exposed to JP-8 indicated that JP-8 exposure for 1 hr per day, 2 times per wk for 9 months may produce neurotoxic effects. In a self-assessment questionnaire, JP-8-exposed Air Force personnel reported more headaches, dizziness, trouble concentrating, balance problems, walking difficulties, forgetfulness, and trouble with gripping objects than an unexposed (control) group. In that study, JP-8-exposed Air Force personnel also showed lower performance than a control group on several neurobehavioral tests and disturbances of balance and altered eye-blink conditioning response. The lack of exposure information makes it difficult to determine the extent of the health risk.

Animal studies have investigated the effects of several jet fuels on a number of neurobehavioral end points. Several studies showed neurobehavioral effects in F344 and Sprague-Dawley rats exposed to JP-8 and JP-5 vapors at concentrations of about 1,000 mg/m^3 for 6 hr per day, 5 days per week for 6 wk or to JP-8 aerosols at concentrations of 1,059 mg/m^3 for 1 hr per day, 5 days per week for 4 wk. No dose-response relationships were demonstrated in the studies. Furthermore, the relevance of the observed neurobehavioral effects to humans is not known, and these positive findings need to be validated against other well-established neurotoxicity end points. However, the findings provide an indication that the interim PEL of 350 mg/m^3 might be too high to be protective of human health.

The subcommittee recommends additional research to measure ambient and breathing-zone concentrations of JP-8 and its constituents (such as naphthalene and toluene) and to determine body burden through assays of biologic samples for JP-8 constituents and metabolites. The findings should be correlated with acute and chronic symptoms and signs experienced by JP-8-exposed people. Preliminary positive findings reported in two neurologic tests (eye-blink and postoral-sway tests) conducted as part of a recent Air Force human study should be validated with standard neurologic tests.

The subcommittee also recommends studies in experimental animals to examine the potential neurotoxic effects of JP-8. Specifically, the subcommittee recommends that neurologic (histologic, physiologic, and behavioral) measures be included in inhalation-toxicity tests with JP-8 vapors and mixtures of vapors and aerosols. Because the composition of JP-8 varies from batch to batch, scientists with expertise in petroleum toxicology should be consulted to design the best approach for testing the neurotoxicity of JP-8 (e.g., testing JP-8 samples at the extremes of their composition ranges or testing JP-8 samples so that the concentrations of component classes can be correlated with toxic end points).

REFERENCES

Andrews, L.S., and R. Snyder. 1986. Toxic effects of solvents and vapors. Pp. 636-668 in Casarett and Doull's Toxicology: The Basic Science of Poisons, 3rd Ed.,

C.D. Klassen, M.O. Amdur, and J. Doull, eds. New York: Macmillan.

Anger, W.K., and D. Storzbach. 2001. Results and discussion -neurobehavioral - interim report. Pp. 65-67 in JP8 Final Risk Assessment. The Institute of Environmental and Human Health (TIEHH), Lubbock, TX. August 2001.

Anthony, D.C., K. Boekelheide, and D.G. Graham. 1983a. The effect of 3,4-dimethyl substitution on the neurotoxicity of 2,5-hexanedione. I. Accelerated clinical neuropathy is accompanied by more proximal axonal swellings. Toxicol. Appl. Pharmacol. 71(3):362-371.

Anthony, D.C., K. Boekelheide, C.W. Anderson, and D.G. Graham. 1983b. The effect of 3,4-dimethyl substitution on the neurotoxicity of 2,5-hexanedione. II. Dimethyl substitution accelerates pyrrole formation and protein crosslinking. Toxicol. Appl. Pharmacol. 71(3):372-382.

Baldwin, C.M., F.P. Houston, M.N. Podgornik, R.S. Young, C.A. Barnes, and M.L. Witten. 2001. Effects of aerosol-vapor JP-8 jet fuel on the functional observational battery, and learning and memory in the rat. Arch. Environ. Health 56(3):216-226.

Bekkedal, M.Y.V., S.M. McInturf, G.D. Ritchie, and J. Rossi III. 2001. Eyeblink conditioning response test used to assess performance in JP-8 exposed air force personnel. Pp. 69-71 in JP8 Final Risk Assessment. The Institute of Environmental and Human Health (TIEHH), Lubbock, TX. August 2001.

Bhattacharya, A. 2001. Postural balance measurements. Risk assessment of acute exposure to jet fuel. Pp. 72-75 in JP8 Final Risk Assessment. The Institute of Environmental and Human Health (TIEHH), Lubbock, TX. August 2001.

Gibson, R.L., S. Shanklin, and R.L. Warner. 2001a. Health effects comparisons. Pp. 125-129 in JP-8 Final Risk Assessment Report. The Institute of Environmental and Human Health (TIEHH), Lubbock, TX. August 2001.

Gibson, R.L., S. Shanklin, and R.L. Warner. 2001b. Self-reported health status. Pp. 132-139 in JP-8 Final Risk Assessment Report. The Institute of Environmental and Human Health (TIEHH), Lubbock, TX. August 2001.

Graham, D.G., V. Amarnath, W.M. Valentine, S.J. Pyle, and D.C. Anthony. 1995. Pathogenetic studies of hexane and carbon disulfide neurotoxicity. Crit. Rev. Toxicol. 25(2):91-112.

Knave, B., H.E. Persson, J.M. Goldberg, and P. Westerholm. 1976. Long-term Exposure to jet fuel: An investigation on occupationally exposed workers with special reference to the nervous system. Scand. J. Work Environ. Health 2(3):152-164.

Knave, B., B.A. Olson, S. Elofsson, F. Gamberale, A. Isaksson, P. Mindus, H.E. Persson, G. Struwe, A. Wennberg, and P. Westerholm. 1978. Long-term exposure to jet fuel. II. A cross-sectional epidemiologic investigation on occupationally exposed industrial workers with special reference to the nervous system. Scand. J. Work Environ Health. 4(1):19-45.

Knave, B., P. Mindus, and G. Struwe. 1979. Neurasthenic symptoms in workers occupationally exposed to jet fuel. Acta Psychiatr. Scand. 60(1):39-49.

Koschier, F.J. 1999. Toxicity of middle distillates from dermal exposure. Drug Chem. Toxicol. 22(1):155-164.

Marshall, B.E., and H. Wollman. 1985. General anesthetics. Pp. 276-301 in Goodman and Gilman's Pharmacological Basis of Therapeutics, 7th Ed., A.G. Gilman, L.S. Goodman, T.W. Rall, and F. Murad, eds. New York: Macmillan.

Nordholm, A.F., J. Rossi III, G.D. Ritchie, S. McInturf, M.E. Hulme, C. McCool, L. Narayanan, K.L. MacMahon, J. Eggers, H.F. Leahy, and R.E. Wolfe. 1999. Repeated exposure of rats to JP-4 vapor induces changes in neurobehavioral capacity and 5HT/5-HIAA levels. J. Toxicol. Environ. Health 56(7):471-499.

NRC (National Research Council). 1996. Permissible Exposure Levels for Selected Military Fuel Vapors. Washington, DC: National Academy Press.

Porter, H.O. 1990. Aviators intoxicated by inhalation of JP-5 fuel vapors. Aviat. Space Environ. Med. 61(7):654-656.

Ritchie, G.D., K.R. Still, W.K. Alexander, A.F. Nordholm, C.L. Wilson, J. Rossi III and D.R. Mattie. 2001a. A review of the neurotoxicity risk of selected hydrocarbon fuels. J. Toxicol. Environ. Health Part B Crit. Rev. 4(3):223-312.

Ritchie, G.D., J. Rossi III, A.F. Nordholm, K.R. Still, R.L. Carpenter, G.R. Wenger, and D.W. Wright. 2001b. Effects of repeated exposure to JP-8 jet fuel vapor on learning of simple and difficult operant tasks by rats. J. Toxicol. Environ. Health Part A 64(5):385-415.

Ritchie, G.D., G.R. Wenger, M.Y.V. Bekkedal, R.L. Carpenter, D. Wright, A.F. Nordholm, and J. Rossi III. 2000. Long-term effects of repeated exposure to JP-8 fuel vapor on higher cognitive capacity in rats. Soc. Neurosci. Abstr. 26:263.

Rossi, J., A.F. Nordholm, R.L. Carpenter, G.D. Ritchie, and W. Malcolm. 2001. Effects of repeated exposure of rats to JP-5 or JP-8 jet fuel vapor on neurobehavioral capacity and neurotransmitter levels. J. Toxicol. Environ. Health A 63(6):397-428.

6

Effects of Jet-Propulsion Fuel 8 on the Immune System

This chapter summarizes the studies that investigated the potential toxicity of jet-propulsion fuel 8 (JP-8), related fuels, and kerosene in humans and experimental animals. The 1996 National Research Council report *Permissible Exposure Levels for Selected Military Fuel Vapors* did not specifically consider the immunotoxic effects of JP-8 or related fuels. No earlier studies that addressed the influence of JP-8 exposure on the functional capacity of the immune system to respond to antigenic challenge were available. However, standard histopathologic, hematologic, and clinical chemistry determinations made as part of a standard toxicology and pathology profile data after exposure to several kerosene-based fuels, including JP-5 and JP-8, did not generate concern about immunotoxicity. Immunotoxicity is typically first detected in standard toxicology and pathology studies.

IMMUNOSUPPRESSIVE EFFECTS OF JP-8

The subcommittee reviewed several recent immunotoxicity studies of JP-8 that used immune-function assays (see Table 6-1). However, the methods for those studies largely have not been standardized through interlaboratory com-

TABLE 6-1 Immunosuppressive Effects of JP-8 Exposure in Humans and Experimental Animals

Species or Cell Line	Exposure Concentration	Exposure Duration	Effects	Reference
INHALATION EXPOSURE				
Human	Measurements taken in breathing zones of subjects; median concentration of naphthalene, 1.9 μg/m^3 (low-exposure group), 447 μg/m^3 (high-exposure group); median concentration of benzene, 3.1 μg/m^3 (low-exposure group), 242 μg/m^3 (high-exposure group)	High-exposure group, persistent exposure to JP-8(defined as at least 1 hr twice per wk for at least 9 mo); low-exposure group, no significant exposure to jet fuel or solvents	High-exposure group had higher white-cell counts than low-exposure group; there were increased numbers of neutrophils and monocytes but no differences in total lymphocytes, T cells, NK cells, B cells; white cell, neutrophil, and monocyte counts in high-exposure group did not exceed range of normal values	Rhodes et al. 2001[a]
Human	Exposed group (5,706 people) had potential occupational exposure to JP-8. Control group (5,706 people) did not work in occupations in which exposure to JP-8 would occur	Not reported	Health-event analysis did not find differences in immunologic measures (such as infections) between exposed and control groups	Gibson et al. 2001a[a]
Human	Measurements taken in breathing zones of subjects; median concentration of naphthalene, 1.9 μg/m^3 (low-exposure group), 10.4 μg/m^3 (moderate-exposure group),	High- and moderate-exposure groups, persistent exposure to JP-8; low exposure group, no significant	Analysis of self-assessment questionnaire did not report differences among groups in immunologic-related illnesses	Gibson et al. 2001b[a]

	447 $\mu g/m^3$ (high-exposure group); median concentration of benzene, 3.1 $\mu g/m^3$ (low-exposure group), 7.45 $\mu g/m^3$ (moderate-exposure group), 242 $\mu g/m^3$ (high-exposure group)	exposure to jet fuel or solvents		
F344 rat and C57Bl/6 mouse	500 and 1,000 mg/m^3	90 days continuously, followed by recovery until approximately 24 mo of age	No treatment-related changes in spleen weight or hematology were observed	Mattie et al. 1991
C57BL/6 mouse	100, 250, 500, 1,000, 2,500 mg/m^3 (aerosol)	1 hr/day for 7 days (nose-only)	Exposure at 100 mg/m^3 led to decreased cellularity of thymus; exposure at 500 mg/m^3 led to decreased spleen weight, cellularity; splenic T cells, B cells, macrophages were also affected by exposure at 100 and 500 mg/m^3; splenic T cells, B cells, macrophages were also decreased in JP-8-exposed mice; bone marrow cellularity increased after exposure at 100, 250 mg/m^3 but decreased after exposure at higher concentrations; exposure at 250 mg/m^3 led to reduced spleen cell proliferation responses in vitro after stimulation with Con A or Con A + IL-2	Harris et al. 1997a

(Continued)

TABLE 6-1 *Continued*

Species or Cell Line	Exposure Concentration	Exposure Duration	Effects	Reference
C57BL/6 mouse	1,000, 2,500 mg/m^3 (aerosol)	1 hr/day for 7 days (nose-only)	In mice exposed at both doses, spleen cellularity and spleen cell proliferation persisted for more than 21 days; spleen cells in mice exposed at 1,000 mg/m^3 were suppressed in ability to mediate NK activity, LAK responses, CTL responses	Harris et al. 1997b
C57BL/6 Mouse	250-2,500 mg/m^3 JP-8 aerosol + 1 μM or 1nM substance P aerosol	1 hr/day for 7 days (nose-only)	Substance P administration prevented loss of spleen and thymus cellularity after exposure to JP-8; it also partially restored proliferative response of spleen cells to Con A + IL-2	Harris et al. 1997c
C57BL/6 Mouse	1,000 mg/m^3 (aerosol)	1 hr/day for 7 days (nose-only)	Mice showed significantly decreased NK cell function, significantly suppressed generation of LAK cell activity, suppressed generation of CTL cells from precursor T cells, inhibited helper T cell activity	Harris et al. 2000
DERMAL EXPOSURE				
C3H/HeN Mouse	50, 250-300 μL	5 days (50 μL), single dose (250-300 μL)	Induction of contact hypersensitivity was impaired in dose-dependent manner regardless of whether contact	Ullrich 1999

			allergen was applied directly to treated skin or at distant unrelated site; generation of classic delayed hypersensitivity reaction to *Borellia burgdorferi* (bacterial antigen) injected into subcutaneous space was suppressed by dermal application of JP-8 at distant site; ability of splenic T lymphocytes from JP-8-treated mice to proliferate in response to plate-bound monoclonal anti-CD3 was significantly suppressed; IL-10 was found in the serum of JP-8-exposed mice	
C3H/HeN mouse	50-300 μL undiluted or diluted in acetone	Single dose	Splenic T-cells were cultured in vitro with antibody T-cell receptor; T cells from JP-8-exposed mice had reduced proliferative response; T-cell-dependent antibody responses to KLH antigen injected in Freund's adjuvant were not altered by exposure to JP-8	Ullrich and Lyons 2000
NHEK	80-200 μg/mL diluted in absolute ethanol	Single dose	JP-8 induced necrosis and cell death in human keratinocytes in vitro	Rosenthal et al. 2001
NHEK	0.1%	Single dose for 24 hr	JP-8 increased production of proinflammatory cytokines TNFα and IL-8	Allen et al. 2000

(Continued)

TABLE 6-1 *Continued*

Species or Cell Line	Exposure Concentration	Exposure Duration	Effects	Reference
F344 rat	0.25 mL	Single dose	IL-1α, iNOS expression were induced in isolated skin samples	Kabbur et al. 2001
ORAL EXPOSURE				
B6C3F1 mouse	1,000, 2,000 mg/kg per day	Administered to pregnant mice on days 6-15 of gestation	Significant suppression of PFC response in offspring when tested at age of 8 wk	Keil et al. 2001
B6C3F1 and DBA/2 mouse	1,000, 2,000 mg/kg per day	1 dose/day for 7 or 14 days	Significant immunologic alterations in thymic weight and antibody PFC response to SRBC	Dudley et al. 2001

[a]Background information about these studies can be found in Appendix B.

Abbreviations: Con A, concanavalin A; CTL, cytotoxic T lymphocytes; IL, interleukin; iNOS, inducible nitric oxide synthase; KLH, keyhole-limpet hemocyanin; LAK, lymphokine-activated killer cell; NHEK, normal human epidermal keratinocyte; NK, natural killer cell; PFC, plaque-forming cell; SRBC, sheep red blood cell; TNF, tumor necrosis factor.

parisons or validated for predictability (Luster et al. 1988, 1992). Several of the approaches used in recent studies would more typically be conducted as mechanistic studies, assuming that significant immunotoxicity was found in standardized toxicology and pathology studies. The potential significance of these recent findings is discussed below; however, it should be noted that the subcommittee expressed concern about the adequacy of exposure characterization and assay validation for many of the studies.

Inhalation Exposure

Carpenter et al. (1976) reported no statistically significant or treatment-related microscopic or histopathologic changes in the spleen of rats or dogs exposed to deodorized kerosene at up to 100 mg/m^3 for 6 hr/day, 5 days/wk for 13 wk. Mattie et al. (1991) exposed Fisher 344 rats and C57Bl/6 mice of both sexes to JP-8 vapors at 0, 500, and 1,000 mg/m^3 on a continuous basis for 90 days, followed by recovery until the age of about 24 months. Fifteen rats and 25 mice per dose group were sacrificed at exposure termination and necropsied, and there were interim sacrifices and necropsies. No statistically significant differences in spleen weight or hematologic measures were observed between exposed and control rats at any time. At terminal sacrifice, female rats showed increased hematopoiesis in the spleen that was dose-dependent but minimal to mild and not considered treatment-related. In mice, no significant clinical signs of JP-8 toxicity were noted. An increased incidence of deaths in treated mice appeared to be due to an increased incidence of necrotic dermatitis due to fighting. Nine months after termination of exposure, pathologic findings were limited to an increased incidence of inflammatory skin lesions and splenic hematopoiesis in male mice; these findings were not considered to be treatment-related. At 24 months after termination of exposure, histopathologic findings were minimal. Histopathologic findings at exposure termination were minimal. Nine months after exposure, pathologic findings were limited to increased incidence of inflammatory skin lesions and splenic hematopoiesis in male mice; neither effect was considered treatment-related.

In contrast with the results by Mattie et al. (1991), Harris et al. (1997a,b,c, 2000) reported significant immunopathologic effects in C57Bl/6 mice exposed nose-only to aerosolized JP-8 (with a median mass aerodynamic diameter [MMAD] of 1.7-1.9 μm; M. Witten, University of Arizona, personal communication, 2002). Reported exposure at 100 mg/m^3 for 1 hr/day for 7 days resulted in decreased cellularity of the thymus gland while exposure at 500 mg/m^3 for 1 hr/day for 7 days resulted in decreased spleen weight and cellularity. Splenic T cells, B cells, and macrophages were affected to a similar degree. Bone marrow cellularity increased at 100 and 250 mg/m^3 and then decreased at higher concentrations. Because body-weight data were not reported, it is unclear whether the changes in lymphoid tissue cellularity represent general toxicity or specific changes in lymphoid tissue. A minimal exposure at 250 mg/m^3 1 hr/day for 7 days also resulted in reduced spleen cell proliferation responses in vitro after stimulation with concanavalin A (Con A) or Con A and interleukin-2 (IL-2) (Harris et al. 1997a). At 1,000 or 2,500 mg/m^3, changes in spleen cellularity and spleen cell proliferation persisted for more than 21 days after the last exposure (Harris et al. 1997b). The ability of spleen cells to mediate natural killer-cell activity, lymphokine-activated killer-cell responses, or cytotoxic T-lymphocyte responses was also suppressed when cells were obtained from mice exposed to JP-8 at 1,000 mg/m^3. Those results suggest that inhalation exposure to aerosolized JP-8 suppressed many of the functions of isolated spleen cells in culture that are considered to reflect the status of the immune system.

No published studies have shown that in vivo immune responses or resistance to infectious disease challenges were altered in JP-8-exposed people. However, preliminary data suggest that JP-8-exposed mice do not regulate the growth of pulmonary B16 melanoma cells as well as control mice (D.T. Harris, University of Arizona, personal communication, 2001) and experience a higher mortality after nasal challenge with Hong Kong influenza virus (M. Witten, University of Arizona, personal communication, 2001). Because inhalation of aerosolized JP-8 can cause local irritation and overt injury to the lung (Robledo and Witten 1999), alterations in B16 melanoma metastases could reflect alterations in the initial deposition of intravenously injected tumor cells as a result of lung-tissue damage rather than of alterations in immune function.

Physical changes in the lung after exposure to aerosolized JP-8 may also underlie the altered systemic effects on lymphoid tissue. Robledo and Witten (1999) reported that treatment of mice with substance P, a neurokinin receptor agonist, protected the lungs from the damaging effects of aerosolized JP-8, including increased permeability, epithelial necrosis, and perivascular edema. Substance P administration was also reported to prevent the loss of spleen and thymus cellularity after JP-8 exposure (250-2,500 mg/m^3) and to partially restore the proliferative response of spleen cells to Con A + IL-2 (Harris et al. 1997c).

The results of the studies by Harris and colleagues (1997 a,b,c; 2000) raise concerns about the immunotoxic potential of JP-8 exposure. However, there are also questions as to why those studies showed such profound changes in lymphoid tissues when prior studies that examined the effects of vaporized JP-8 failed to show such effects. Specifically, the subcommittee suspects that the actual exposures in Harris et al. (1997 a,b,c; 2000) were underreported. JP-8 concentrations were not assessed in the aerosol, blood, or tissue, and this could lead to erroneous assumptions regarding exposure concentrations. However, even if the actual concentrations were 10 times as high (e.g., exposure was at a concentration of 1,000 mg/m^3), the observation of positive effects from a short duration (1 hr/day for 7 days) at that concentration might yield a safe exposure level of less than 350 mg/m^3 (assuming the application of commonly used uncertainty factors). It is also likely that exposure to aerosolized JP-8 is more toxic than exposure to vaporized JP-8. In addition, the nose-only exposure protocol may have concentrated the JP-8 aerosol and led to an increase in cell membrane damage, or it may have induced stress in the animals, compared with whole-body exposure.

A comparison of changes in white-blood-cell counts after JP-8 exposure may be revealing in that white-blood-cell counts in military personnel exposed to JP-8 have been measured (Rhodes et al. 2001). In mice, white-blood-cell counts were decreased after exposure to JP-8 at lower concentrations (i.e., 100 and 250 mg/m^3) and increased at higher concentrations (Harris et al. 1997a). Differential analysis revealed concentration-dependent neutrophilia. In addition, JP-8 exposure was associated with a concentration-dependent reduction in T cells and macrophages but no effect on B cells. At the highest exposure concentration (2,500 mg/m^3) white-blood-cell counts were reduced to numbers insufficient for analysis. In military personnel exposed to JP-8, tank-entry workers, considered to be among the most highly exposed population, were found to have higher white-blood-cell counts than a low-exposure group (Rhodes et al. 2001). Differential analyses revealed increased numbers of neutrophils and monocytes but no differences in total numbers of lymphocytes, T cells, NK cells, or B cells. Those findings suggest that JP-8 exposure might induce an inflammatory response in humans but do not corroborate the decrease in immune-cell numbers seen in mice. It is important to note that the increase in white-blood-cells, neutrophils, and monocytes in the military personnel exposed to JP-8 did not exceed the normal ranges. A previous pilot study by Olsen et al. (1998) found no difference in total white-blood-cell and differential counts among Air Force personnel before and 18 months after the Air Force converted to JP-8.

The functional status of the immune system has not been evaluated in military personnel exposed to JP-8. If impairment of immune function was induced, one would expect to see an increased incidence and severity of infectious disease in highly exposed workers. However, a health-event analysis of outpatient medical records conducted by the Air Force (Gibson et al. 2001a) found no differences in health-seeking events between fuel-cell workers and other base personnel. In a related study, self-reported prevalence of illness did not differ between moderate and high-exposure groups and a low-exposure group (Gibson et al. 2001b); a limitation of this study was that the questionnaire used addressed only ear infections, and the incidence of colds or flu might have been more relevant.

Dermal Exposure

Ullrich (1999) reported that dermal exposure of C3H/HeN mice to JP-8 in multiple small doses (50 μL/day for 4-5 days) or in larger single doses (300 μL) resulted in local and systemic effects on immune responses. Contact- and delayed-hypersensitivity responses were suppressed by JP-8 exposure. That induction of a contact-hypersensitivity response was reduced when a contact allergen was applied directly to JP-8-treated skin or at a distant site indicates both local and systemic immune suppression. Similarly, the delayed-hypersensitivity response to a bacterial antigen injected subcutaneously was suppressed by dermal JP-8 exposure. When splenic T cells were stimulated to divide in vitro by cross-linking the T-cell receptors, T cells from JP-8-exposed mice showed reduced proliferative response. In contrast, the antibody response to keyhole limpet hemocyanin in adjuvant was not altered by JP-8 exposure (Ullrich and Lyons 2000). Serum concentration of IL-10, a cytokine that suppresses some T-cell functions, was increased within 48 hr after JP-8 exposure (Ullrich 1999). Furthermore, neutralization of IL-10, administration of IL-12 (to bypass IL-10 effects), or blocking of prostaglandin E2 production abrogated the immunotoxic effects of JP-8. The authors hypothesize that IL-10 and prostaglandin E2 are produced as a result of damage to keratinocytes and are released systemically and induce immunosuppression by JP-8 that acts selectively on cell-mediated immune responses. JP-8 has been shown to induce necrosis and cell death in human keratinocytes in vitro (Rosenthal et al. 2001) and to increase the production of proinflammatory cytokines TNFα and IL-8 (Allen et al. 2000). Dermal exposure of rats also induced IL-1α and inducible nitric oxide synthetase expression in isolated skin samples (Kabbur et al. 2001).

The immunosuppressive effects of dermal JP-8 were dose-dependent: 50 μL for 1-3 days was not significantly suppressive, nor were single doses smaller than 300 μL. The effects of JP-8 were also time-dependent: T-cell proliferation was suppressed within 3-4 days after a single exposure and lasted for about 3 wk. The human dermal dose equivalent to the threshold 300-μL dose in the mouse was calculated by the authors to be 100 mL. Those results raise concern about potential health effects of prolonged or repeated dermal exposure of military personnel to JP-8.

Oral Exposure

No changes in spleen weight or splenic histologic findings were observed after a single oral dose of kerosene at 12,000 mg/kg or of deodorized kerosene at 12,150 mg/kg in rats. Parker et al. (1981) reported a decrease in white cells in rats after a single oral dose of JP-5 at 18,912 mg/kg and an increase in red cells, postulated to be due to hemoconcentration related to dehydration.

Mattie et al. (1995) exposed rats to JP-8 in the diet for 90 days at 0, 750, 1,500, or 3,000 mg/kg. Circulating neutrophil counts increased and lymphocytes decreased. There were no histologic changes in the spleen or lymph nodes, but relative spleen weight was increased at the highest exposure concentration. Mice exposed to JP-8 at 1,000 or 2,000 mg/kg per day for 7 or 14 days via oral gavage had significant immunologic alterations, including decreases in thymic weight and antibody plaque-forming cell response to sheep red-blood cells (Dudley et al. 2001). The suppression of the plaque-forming cell response occurred in the absence of changes in spleen cellularity. The absence of significant differences in resistance to *Listeria* infection or growth of B16 melanoma cells suggests selective effects of JP-8 on humoral immunity.

The selectivity of oral exposure for suppressing humoral rather than cell-mediated immune function is the opposite of what was observed after dermal exposure to JP-8. Because JP-8 is irritating to the gastrointestinal tract and because the oral route is not considered to be relevant to routine occupational exposures, those data were not considered relevant to the subcommittee's charge.

ALLERGIC POTENTIAL OF JP-8

If components of JP-8 are seen as foreign by the immune system, JP-8 exposure could produce an immune response that leads to allergic response. Symptoms depend on the route of exposure: contact dermatitis after dermal exposure, rhinitis and asthma after inhalation, and vomiting or diarrhea after ingestion. If JP-8 components are systemic sensitizers, anaphylaxis, a life-threatening systemic allergic reaction, could occur following subsequent JP-8 exposure. Except for anaphylaxis, similar symptoms can result from a nonspecific inflammatory response to irritants that does not involve sensitization of the immune system.

The murine local lymph node assay is one predictor of the skin-sensitization potential of chemicals (Basketter et al. 1996). When JP-8 was tested in the assay with CBA/Ca mice, a strain that shows increased responsiveness to contact allergens (Kimber and Weisenberger 1989), increased lymphocyte proliferation was observed, with a stimulation index of 3.17. An index greater than 3 is considered evidence of skin sensitization; apparently JP-8 was a weak skin sensitizer. Exposure to Jet A and JP-8 + 100 also increased lymphocyte proliferation, but the indexes were less than 3. Those results suggest that the additives in JP-8 + 100 may reduce the sensitization potential of JP-8 (Kanikkannan et al. 2000). Kinkead et al. (1992a) reported that topical application of JP-8 also showed weak skin sensitization in guinea pigs. Studies with other jet fuels have indicated only weak skin sensitization if any (Cowan and Jenkins 1981; Schultz et al. 1981; Kinkead et al. 1992b). No studies that evaluated sensitization after inhalation of JP-8 were found. Allergic sensitization of humans to JP-8 or other jet fuels has not been reported.

AUTOIMMUNE EFFECTS OF JP-8

No studies that addressed the effects of JP-8 exposure on development or exacerbation of autoimmune disease were found.

CONCLUSIONS AND RECOMMENDATIONS

No histopathologic effects related to the immune system were found in F344 rats and C57BL/6 mice exposed continuously to JP-8 vapors at concentrations up to 1,000 mg/m^3 for 90 days. No additional studies that tested the toxicity of JP-8 vapors in experimental animals were located.

Harris et al. reported that inhalation exposure of C57BL/6 mice to JP-8 aerosols at a concentration of 100 mg/m^3 for 1 hr/day for 7 days led to decreased cellularity of the thymus, exposure at 500 mg/m^3 for 1 hr/day for 7 days led to decreased spleen weight and cellularity, and exposure at 1,000 mg/m^3 for 1 hr/day for 7 days led to decreased ability of spleen cells to mediate several immune responses. Those studies raise concern about the potential of JP-8 to cause immunotoxicity. The subcommittee reviewed the methods used to generate the exposure atmospheres in the studies by Harris et al. and suspects that the total JP-8 concentration in the atmosphere may have been underreported. However, even if the actual concentration was 10 times as high as the lowest concentration at which effects were observed (100 mg/m^3) (i.e., if exposure was at a concentration of 1,000 mg/m^3), the observation of positive effects from a short exposure duration (1 hr/day for 7 days) at that concentration leads the subcommittee to conclude that the interim permissible exposure level of 350 mg/m^3 might be too high to be protective of human health (assuming the application of commonly used uncertainty factors).

Dermal exposure of mice to JP-8 in multiple small doses (50 μL/day for 4-5 days) or in larger single doses (300 μL) resulted in local and systemic effects on the immune system (e.g., suppressed contact- and delayed-hypersensitivity responses).

The subcommittee recommends that experimental animal studies examining the immunotoxicity of JP-8 via the inhalation route be conducted with careful control of vapor and aerosol concentrations in the atmosphere and with consideration of appropriate controls. The studies need to be designed in collaboration with scientists who are knowledgeable about aerosol generation, aerosol physics, and quantification of vapors and aerosols to ensure accurate characterization of the exposure atmospheres. Because the composition of JP-8 varies from batch to batch, scientists with expertise in petroleum toxicology should be consulted to design the best approach for testing the immunotoxicity of JP-8 (e.g., testing JP-8 samples at the extremes of their composition ranges or testing JP-8 samples so that the concentrations of component classes can be correlated with toxic end points).

The subcommittee recommends that human blood samples from JP-8-exposed persons be assayed for indicators of immunotoxicity to determine whether effects in experimental animals are observed in humans.

Furthermore, the subcommittee recommends that military personnel avoid direct, prolonged skin contact with JP-8.

REFERENCES

Allen, D.C., J.E. Riviere, and N.A. Monteiro-Riviere. 2000. Identification of early biomarkers of inflammation produced by keratinocytes exposed to jet fuels Jet A, JP-8, and JP-8(100). J. Biochem. Mol. Toxicol. 14(5):231-237.

Basketter, D.A., G.F. Gerberick, I. Kimber, and S.E. Loveless. 1996. The local lymph node assay: A viable alternative to currently accepted skin sensitization tests. Food Chem. Toxicol. 34(10):985-997.

Carpenter, C.P., D.L. Geary Jr., R.C. Myers, D.J. Nachreiner, L.J. Sullivan, and J.M. King. 1976. Petroleum hydrocarbon toxicity studies. XI. Animal and human response to vapors of deodorized kerosene. Toxicol. Appl. Pharmacol. 36(3):443-456.

Cowan, M.J., and L.J. Jenkins. 1981. U.S. Navy toxicity study of shale and petroleum JP-5 aviation fuel and diesel fuel marine. Pp. 129-140 in Health Effects Investigation of Oil Shale Development, M.R. Guerin, W.H. Griest, and D.L. Coffin, eds. Ann Arbor, MI: Ann Arbor Science.

Dudley, A.C., M.M. Peden-Adams, J. EuDaly, R.S. Pollenz, and D.E. Keil. 2001. An aryl hydrocarbon receptor independent mechanism of JP-8 jet fuel immunotoxicity in Ah-responsive and Ah-nonresponsive mice. Toxicol. Sci. 59(2):251-259.

Gibson, R.L., S. Shanklin, and R.L. Warner. 2001a. Health effects comparisons. Pp. 125-129 in JP-8 Final Risk Assessment Report. The Institute of Environmental and Human Health (TIEHH), Lubbock, TX. August 2001.

Gibson, R.L., S. Shanklin, and R.L. Warner. 2001b. Self-reported health status. Pp. 132-139 in JP-8 Final Risk Assessment Report. The Institute of Environmental and Human Health (TIEHH), Lubbock, TX. August 2001.

Harris, D.T., D. Sakiestewa, R.F. Robledo, and M. Witten. 1997a. Immunotoxicological effects of JP-8 jet fuel exposure. Toxicol. Ind. Health 13(1):43-55.

Harris, D.T., D. Sakiestewa, R.F. Robledo, and M. Witten. 1997b. Short-term exposure to JP-8 jet fuel results in long-term immunotoxicity. Toxicol. Ind. Health 13(5): 559-570.

Harris, D.T., D. Sakiestewa, R.F. Robledo, and M. Witten. 1997c. Protection from JP-8 jet fuel induced immunotoxicity by administration of aerosolized substance P. Toxicol. Ind. Health 13(5):571-588.

Harris, D.T., D. Sakiestewa, R.F. Robledo, and M. Witten. 2000. Effects of short-term JP-8 jet fuel exposure on cell-mediated immunity. Toxicol. Ind. Health 16(2):78-84.

Kabbur, M.B., J.V. Rogers, P.G. Gunasekar, C.M. Garrett, K.T. Geiss, W.W. Brinkley, and J.N. McDougal. 2001. Effect of JP-8 jet fuel on molecular and biological parameters related to acute skin irritation. Toxicol. Appl. Pharmacol. 175(1):83-88.

Kanikkannan, N., T. Jackson, M. Sudhan Shaik, and M. Singh. 2000. Evaluation of skin sensitization potential of jet fuels by murine local lymph node assay. Toxicol. Lett. 116(1-2):165-170.

Keil, D.E., D.A. Warren, M.M. Peden-Adams, and J. EuDaly. 2001. The effects of JP-8 on immune function and thyroid hormone levels in B6C3F1 mice exposed in utero. Toxicologist 60(1):218.

Kimber, I., and C. Weisenberger. 1989. A murine local lymph node assay for the identification of contact allergens. Assay development and results of an initial validation study. Arch. Toxicol. 63(4):274-282.

Kinkead, E.R., S.A. Salins, and R.E. Wolfe. 1992a. Acute irritation and sensitization potential of JP-8 jet fuel. J. Am. Coll. Toxicol. 11(6):700.

Kinkead, E.R., R.E. Wolfe, and S.A. Salins. 1992b. Acute irritation and sensitization potential of shale-derived JP-5 jet fuel. J. Am. Coll. Toxicol. 11(6):705.

Luster, M.I., A.E. Munson, P.T. Thomas, M.P. Holsapple, J.D. Fenters, K.L. White Jr., L.D. Lauer, D.R. Germolec, G.J. Rosenthal, and J.H. Dean. 1988. Development of a testing battery to assess chemical-induced immunotoxicity: National Toxicology Program?s guidelines for immunotoxicity evaluation in mice. Fundam. Appl. Toxicol. 10(1):2-19.

Luster, M.I., C. Portier, D.G. Pait, K.L. White Jr., C. Gennings, A.E. Munson, and G.J. Rosenthal. 1992. Risk assessment in immunotoxicology. I. Sensitivity and predictability of immune tests. Fundam. Appl. Toxicol. 18(2):200-210.

Mattie, D.R., C.L. Alden, T.K. Newell, C.L. Gaworski, and C.D. Flemming. 1991. A 90-day continuous vapor inhalation toxicity study of JP-8 jet fuel followed by 20 or 21 months of recovery in Fischer 344 rats and C57BL/6 mice. Toxicol. Pathol. 19(2):77-87.

Mattie, D.R., G.B. Marit, C.D. Flemming, and J.R. Cooper. 1995. The effects of JP-8 jet fuel on male Sprague-Dawley rats after a 90-day exposure by oral gavage. Toxicol. Ind. Health 11(4):423-435.

NRC (National Research Council). 1996. Permissible Exposure Levels for Selected Military Fuel Vapors. Washington, DC: National Academy Press.

Olsen, D.M., D.R. Mattie, W.D. Gould, F. Witzmann, M. Ledbetter, G.K. Lemasters, and J.H. Yin. 1998. Pilot Study of Occupational Assessment of Air Force Personnel Exposure to Jet Fuel Before and After Conversion to JP-8. AFRL-HE-WP-TR-1998-0107. Air Force Research Laboratory, Operational Toxicology Branch, Wright-Patterson AFB, OH. 43pp.

Parker, G.A., V. Bogo, and R.W. Young. 1981. Acute toxicity of conventional versus shale-derived JP5 jet fuel: Light microscopic, hematologic, and serum chemistry studies. Toxicol. Appl. Pharmacol. 57(3):302-317.

Rhodes, A.G., G.K. LeMasters, J.E. Lockey, J.W. Smith, J.H. Yiin, R. Gibson, and S. Rappaport. 2001. The effects of JP8 jet fuel on immune cell counts of tank entry workers. Pp. 100-120 in JP-8 Final Risk Assessment Report. The Institute of Environmental and Human Health (TIEHH), Lubbock, TX. August 2001.

Robledo, R.F., and M.L. Witten. 1999. NK1-receptor activation prevents hydrocarbon induced lung injury in mice. Am. J. Physiol. 276(2 Pt 1):L229-L238.

Rosenthal, D.S., C.M. Simbulan-Rosenthal, W.F. Liu, B.A. Stoica, and M.E. Smulson. 2001. Mechanisms of JP-8 jet fuel cell toxicity. II. Induction of necrosis in skin fibroblasts and keratinocytes and modulation of levels of Bcl-2 family members. Toxicol. Appl. Pharmcol. 171(2):107-116.

Schultz, T.W., H. Witschi, L.H. Smith, W.M. Haschek, J.M. Holland, J.L. Epler, R.M. Fry, T.K. Rao, F.W. Larimer, and J.N. Dumont. 1981. Health Effects Research in Oil Shale Development. ORNL/TM-8034. Oak Ridge, TN: Oak Ridge National Laboratory. 61pp.

Ullrich, S.E. 1999. Dermal application of JP-8 jet fuel induces immune suppression. Toxicol. Sci. 52(1):61-67.

Ullrich, S.E., and H.J. Lyons. 2000. Mechanisms involved in the immunotoxicity induced by dermal application of JP-8 jet fuel. Toxicol. Sci. 58(2):290-298.

7

Effects of Jet-Propulsion Fuel 8 on the Liver

This chapter summarizes the findings on potential hepatic toxicity of jet-propulsion fuel 8 (JP-8) and related fuels presented in the National Research Council report *Permissible Exposure Levels for Selected Military Fuel Vapors* (NRC 1996) and reviews additional studies, most of which were completed after the 1996 report was published. The subcommittee uses that information to assess the potential toxic effects of JP-8 on the human liver.

SUMMARY OF STUDIES DISCUSSED IN THE 1996 NATIONAL RESEARCH COUNCIL REPORT

The National Research Council Subcommittee on Permissible Exposure Levels for Military Fuel Vapors reviewed studies concerning potential hepatic changes associated with exposure to the vapors of JP-8, JP-4, JP5, or diesel fuel marine (DFM) on the liver (NRC 1996).

One study examined the effects of JP-4 on the liver in humans. Dossing et al. (1985) reported that fuel-filling attendants exposed to JP-4 at an average of 31 mg/m^3 for a mean of 6.4 years had a significantly faster antipyrine clearance (68 mL/min) than an referent population of office workers (58 mL/min).

No marked differences were found in serum aspartate aminotransferase and alkaline phosphatase activity between the two groups. No studies were available that report the effects of JP-8, JP-5, or DFM vapors on the liver in humans.

Studies in rats and mice had examined the toxic effects of JP-8 on the liver (MacEwen and Vernot 1983, 1984,1985). In subchronic inhalation studies, male and female F344 rats (10 of each) and male and female C57BL/6 mice (10 of each) were continuously exposed to JP-8 vapor at 500 or 1,000 mg/m^3 for 90 days. Some groups of animals were killed immediately after the 90-day exposure, and others 2 wk, 2 months (mo), 9 mo, or 21 mo after the exposure. Immediately after exposure ceased, male rats showed increases in liver weights and liver:body weight ratios at 1,000 mg/m^3, decreases in serum glutamic-pyruvic transaminase (SGPT) activity at 500 and 1,000 mg/m^3, and decreases in alkaline phosphatase activity at 1,000 mg/m^3; and female rats showed increases in liver weights and liver:body weight ratios at 500 and 1,000 mg/m^3, increases in alkaline phosphatase activity at 1,000 mg/m^3, and decreases in SGPT activity at 500 and 1,000 mg/m^3. Nine months after exposure, male rats showed decreases in SGPT activity at 500 and 1,000 mg/m^3. Twenty-one months after exposure, male rats showed concentration-related increases in liver:body weight ratios at 500 and 1,000 mg/m^3; and female rats showed decreases in serum glutamic oxaloacetic transaminase (SGOT) activity at 500 mg/m^3 and decreases in SGPT activity at 500 and 1,000 mg/m^3. It should be emphasized that despite the significant changes observed in SGOT and SGPT activities, the alterations were within the normal range and thus not clinically relevant. Support for relevant changes in liver function would necessitate the measurement of liver enzyme functions and histopathologic studies, which were not conducted. No data on mice were presented.

The available data on potential hepatic toxicity associated with subchronic exposure to JP-8 vapor are not definitive, because histopathologic examinations were not performed. The liver weight changes observed in rats might indicate hyperplasia or hypertrophy. Alternatively, the increases in liver:body weight ratios might reflect a loss of body weight in the test animals during the study. It is also possible that JP-8 was offensive to the animals, nauseating them and decreasing their food intake.

Animal studies had also examined the liver effects from dermal or inhalation exposure to JP-4 or JP-5. Mild liver changes were observed in male and female beagles, male and female F344 rats, and male and female C57BL/6 mice exposed to JP-5 vapor continuously at 150 or 750 mg/m^3 for 90 days (MacEwen and Vernot 1978, 1980, 1981, 1982, 1983, 1985; Gaworski et al. 1984). Similar results were reported in beagles, F344 rats, and C57BL/6 mice exposed to JP-4 vapor at 500 or 1,000 mg/m^3 for 90 days (MacEwen and

Vernot 1984); in F344 rats and C57BL/6 mice exposed to JP-4 vapor at 1,000 or 5,000 mg/m^3 for 6 hr/day, 5 days/wk for 12 mo (Bruner et al. 1993; Wall et al. 1990; MacEwen and Vernot 1981, 1982); and in monkeys, dogs, rats, and mice exposed to JP-4 vapor at 2,500 or 5,000 mg/m^3 for 6 hr/day, 5 days/wk for 8 mo (MacNaughton and Uddin 1984).

EFFECTS OF EXPOSURE TO JP-8 IN HUMANS

The effects of acute exposure to JP-8 on the liver in humans were examined in a study recently completed by the U.S. Air Force. The preliminary results of that study are described below and summarized in Table 7-1.

Snawder and Butler (2001) collected venous blood and urine from 107 people working at six Air Force bases (AFB): Davis Monthan AFB, Arizona; Seymour Johnson AFB, North Carolina; Langley AFB, Virginia; Pope, AFB, North Carolina; Little Rock AFB, Arkansas; and Hurlbert Field, Florida. The exposed workers were fuel tank-entry personnel with at least 9 mo of persistent exposure to jet fuel (defined as 1-hr entry, twice a week). The unexposed group consisted of Air Force personnel who routinely had no significant exposure to solvents or fuels. The participants completed questionnaires on job category, exposure, and medical and demographic items. The exclusion criteria for participants were the presence of autoimmune disease, cancer, or diabetes and the use of immune-system altering drugs.

Blood samples were collected before and after shift at each AFB and sent to a National Institute for Occupational Safety and Health (NIOSH) laboratory in Cincinnati, Ohio, for analysis. The markers of liver damage included serum alpha-glutathione *S*-transferase (GST) activity, an index of liver toxicity. In measurement with commercial immunoassay kits, hepatic alpha-GST activity in control and exposed subjects fell within the normal range. Butler et al. (2001) further categorized hepatic alpha-GST in three exposure groups to assess correlation of JP-8 exposure with potential liver toxicity. The high-exposure group consisted of subjects who routinely performed tasks associated with repair of aircraft fuel systems; the moderate-exposure group consisted of subjects who were involved with fuel handling, distribution, recovery, and testing; and the low-exposure group consisted of subjects who did not normally come into contact with jet fuel or solvents. That hepatic alpha-GST activity was not significantly different among those groups indicated a lack of interaction between exposure concentration and genotype, and there was no enzymatic induction. In addition, Butler et al. (2001) measured serum cytochrome P2E1 activity; cytochrome P2E1 is an enzyme involved in benzene metabolism to benzene oxide and phenol. Phenol via cytochrome P2E1

TABLE 7-1 Effects of JP-8 Exposure on the Liver in Humans[a]

Reference	Exposure Concentration	Exposure Duration	Results
Snawder and Butler 2001	Measurements taken in breathing zones of subjects; median concentration of naphthalene, 1.9 μg/m^3 (low-exposure group), 447 μg/m^3 (high-exposure group); median concentration of benzene, 3.1 μg/m^3 (low-exposure group), 242 μg/m^3 (high-exposure group)	High-exposure group had persistent exposure to JP-8 (defined as at least 1 hr twice per wk for 9 mo); low-exposure group had no significant exposure to jet fuel or solvents	Concentrations of serum hepatic alpha-GST activity in study subjects were within normal range
Butler et al. 2001	Measurements taken in breathing zones of subjects; median concentration of naphthalene, 1.9 μg/m^3 (low-exposure group), 10.4 μg/m^3 (moderate-exposure group), 447 μg/m^3 (high-exposure group); median concentration of benzene, 3.1 μg/m^3 (low-exposure group), 7.45 μg/m^3 (moderate-exposure group), 242 μg/m^3 (high-exposure group)	High- and moderate-exposure groups had persistent exposure to JP-8; low exposure group had no significant exposure to jet fuel or solvents	Frequency of CYP2E1 and NQOI genotypes was similar in subjects in all exposure groups; no change in enzymatic activity
Gibson et al. 2001a	Exposed group (5,706 people) had potential occupational exposure to JP-8. Control group (5,706 people) did not work in occupations in which exposure to JP-8 would occur	Not reported	Analysis of medical records showed that subjects in all groups had similar health-care visit rates; no differences were noted among groups in digestive ailments

(Continued)

TABLE 7-1 *Continued*

Reference	Exposure Concentration	Exposure Duration	Results
Gibson et al. 2001b	Measurements taken in breathing zones of subjects; median concentration of naphthalene, 1.9 $\mu g/m^3$ (low-exposure group), 10.4 $\mu g/m^3$ (moderate-exposure group), 447 $\mu g/m^3$ (high-exposure group); median concentration of benzene, 3.1 $\mu g/m^3$ (low-exposure group), 7.45 $\mu g/m^3$ (moderate-exposure group), 242 $\mu g/m^3$ (high-exposure group)	High- and moderate-exposure groups had persistent exposure to JP-8; low-exposure group had no significant exposure to jet fuel or solvents	Analysis of self-assessment questionnaire did not report differences among groups in digestive ailments

[a]Data collected from volunteers (male and female active-duty Air Force personnel) at six Air Force bases in United States. Volunteers were divided into three exposure groups: high, moderate, and low. High-exposure group performed tasks associated with repairing aircraft fuel systems; moderate-exposure group performed tasks associated with fuel handling, distribution, recovery, and testing; and low-exposure group did not routinely come into contact with jet fuel or solvents. Data were collected in morning before subjects went to work and again after they completed their work for that day. Reported results are from preliminary analysis of data. Work referred to in table is part of larger study examining potential human health effects of acute exposure to JP-8. Additional background information can be found in Appendix B.

Abbreviations: GST, glutathione-*S*-transferase; CYP2E1, cytochrome P2E1; NQO1, NAD(P)H quinone oxidoreductase.

is oxidized to hydroquinone and other quinines, including benzoquinones. NAD(P)H quinone oxidoreductase (NQOI) then catalyzes conversion of benzoquinones to less-reactive metabolites. The frequency of the cytochrome P2E1 and NQOI genotypes was similar in subjects regardless of exposure concentration, and there was no change in enzymatic activity, so there was probably no hepatic metabolic induction. Data indicated that those sensitive measures of risk did not detect adverse effects of JP-8 at the assumed exposures on human liver function.

Gibson et al. (2001a) examined the medical records of Air Force personnel occupationally exposed to JP-8 and compared them with records of an unexposed population. The data used were from a population of 5,706 (242 women and 5,464 men) in the exposed group and a population of 5,706 (2,853 men and 2,853 women) randomly chosen from a cohort of 20,244 Air Force personnel who were not occupationally exposed to JP-8. The total number of health-event visits was not markedly different between groups. There was no association between JP-8 exposure and specific neoplasia or digestive ailments. Furthermore, Gibson et al. (2001b) conducted a self-assessment questionnaire on 328 exposed people, categorized into high-, moderate-, and low-exposure groups (as described above). In both men and women, the incidence of digestive ailments was not markedly different between the exposed and referent groups.

EFFECTS OF EXPOSURE TO JP-8 IN EXPERIMENTAL ANIMALS

Several studies have been conducted to examine the potential adverse effects of JP-8 on liver function. Those studies are described below and summarized in Table 7-2.

Parton (1994) subjected male F344 rats to nose-only inhalation exposure to JP-8 aerosols (average particle size was 1.1054 ± 0.2918 microns) at 500 or 1,000 mg/m^3 for 1 hr/day for 7 or 28 days. Weight gain in the 28-day low- and high-dose groups was significantly decreased, but the final body weight was not markedly different between groups. Liver weights were not significantly different. There were no significant alterations in serum aspartate aminotransferase (AST) and alanine aminotransferase (ALT) activities, indicators of hepatic function, and there were no marked changes in the liver histopathologic findings and cytochrome P450 content, a measure of xenobiotic metabolism.

Mattie et al. (1991) exposed male and female F344 rats and male and female C57BL/6 mice to JP-8 vapor at 500 or 1,000 mg/m^3 for 90 days. Only

TABLE 7-2 Effects of Jet Fuel Exposure on the Liver in Experimental Animals

Fuel Type	Species	Exposure Concentration	Exposure Duration	Effects	Reference
JP-8	Male, female F344 rats	500 or 1,000 mg/m^3 (vapor)	90 days continuously	Increased in liver weight and liver:body weight ratio, decreased in SGPT activity in males and females at 500 or 1,000 mg/m^3; decreased alkaline phosphatase activity in males, increased alkaline phosphastase activity in females at 1,000 mg/m^3	MacEwen and Vernot 1983, 1984, 1985
JP-8	Male F344 rats	500 or 1,000 mg/m^3 (aerosol, nose-only)	1 hr/day for 7 or 28 days	Body weight gain in rats exposed for 28 days was significantly decreased; final body weights of exposed animals were similar to those of control animals; liver weights not significantly different between groups; relative liver weight increased in high-dose groups; no significant alterations in AST and ALT activity; no marked changes in liver histopathologic findings and CYP450 content	Parton 1994
JP-8	Male, female F344 rats, C57Bl/6 mice	500 or 1,000 mg/m^3 (vapor)	90 days continuously	Male rats had a statistically significant increase in hepatic basophilic foci. Their presence in the livers of male rats is of uncertain biological significance. No alterations were found in hepatic tissue of female rats or in mice	Mattie et al. 1991

JP-8	Male Sprague-Dawley rats	750, 1,500, or 3,000 mg/kg (gavage)	90 days consecutively	Serum ALT and AST activity increased significantly in all groups, but increase was not dose-related; liver weight similar in all groups; increased relative tissue weight in high-exposure group; liver histologic findings similar in all groups (including control group)	Mattie et al. 1995
JP-8	Male Sprague-Dawley Rats	1,000 mg/m^3 (vapor, whole-body)	6 hr/day, 5 days/wk for 6 wk	Hepatic lamin L83 abundance significantly decreased; lamin L603 abundance increased; total lamin A abundance not significantly altered by JP-8 exposure	Witzman et al. 2000
JP-8	Female B6C3F1, DBA/2 mice	1 or 2 g/kg per day (oral gavage)	7 days	Significantly increased body weights of B6C3F1 mice, but not DBA/2 mice; increased liver:body weight ratios in both strains; no marked change in expression of CYP1A1	Dudley et al. 2001
JP-5	Beagles, F344 rats, C57BL/6 mice	150 or 750 mg/m^3 (vapor)	90 days continuously	Reversible diffuse mild swelling of hepatocytes, decreased SGPT activity, increased liver weight in dogs; mild hepatic hyperplasia, increased hepatocyte vacuolization in rats; fatty changes in hepatocytes, increased hepatocytic vacuolization, increased liver adenomas in mice	MacEwen and Vernot 1978, 1980, 1981, 1982, 1983, 1985; Gaworski et al. 1984

(Continued)

TABLE 7-2 *Continued*

Fuel Type	Species	Exposure Concentration	Exposure Duration	Effects	Reference
JP-4	Male, female F344 rats, C57BL/6 mice	1,000 or 5,000 mg/m^3 (vapor)	6 hr/day, 5 day/wk for 12 mo	Decreased liver weights, liver:body weight ratio, SGPT activity in male rats; decreased SGPT activity, presence of liver nodular hyperplasia in high-dose female rats; decreased incidence of adenomas in male high-dose mice; increased liver inflammatory infiltrates, incidence of hepatocellular adenomas in high-dose female mice	MacEwen and Vernot 1981, 1982; Wall et al. 1990; Bruner et al. 1993
JP-4	Beagle dogs, F344 rats, and C57BL/6 mice	500 or 1,000 mg/m^3 (vapor)	90 days continuously	No effects in dogs; increased liver weight, decreased SGOT and SGPT activity in rats; increased hepatocellular fatty changes in mice	MacEwen and Vernot 1984
JP-4	Monkeys, dogs, rats, mice	2,500 or 5,000 mg/m^3 (vapor)	6 hr/day, 5 days/wk for 8 mo	Increased liver weights; no histopathologic changes	MacNaughton and Uddin 1984
Kerosene	Rat	58 mg/m^3 or 231 mg/m^3 (vapor, possibly some aerosol)	Subchronic (duration not specified)	Decreased blood glucose at 58 mg/m^3; increased blood lactate, pyruvate at 231 mg/m^3	Starek and Vojtisek 1986 as cited in ATSDR 1998

Kerosene	Dogs, rats	100 mg/m^3 (deodorized)	6 hr/day, 5 days/wk, 13 wk	No histopathological changes in the livers of dogs and rats; no liver weight changes in dogs	Carpenter et al. 1976

Abbreviations: ALT, alanine aminotransferase; AST, aspartate aminotransferase; CYP, cytochrome P; GST, glutathione-*S*-transferase; SGOT, serum glutamic oxaloacetic transaminase; SGPT, serum glutamic pyruvic transaminase.

in male rats were hepatic basophilic foci found—in livers of 11, 35, and 31% of control, low-dose, and high-dose groups, respectively. The increase in basophilic foci was statistically significant. Basophilic foci are not reliable predictors of potential hepatic carcinogenicity and their presence in the livers of the male rats is of uncertain biological significance. This finding is similar to the specific hyaline nephropathy found in male rats, which does not have biological relevance for humans (see Chapter 8). The observations that no other hepatic alterations were found in male rats and that no alterations were found in hepatic tissue of female rats or of mice diminish the biologic meaning of hepatic foci changes in male rat liver for humans.

In a study by Mattie et al. (1995), Sprague-Dawley rats were given JP-8 daily for 90 days at 750, 1,500, or 3,000 mg/kg by oral gavage. Serum samples were collected 24 hr before sacrifice. Blood and tissue samples were obtained at sacrifice. With respect to hepatic function, serum ALT and AST activity increased significantly in all three groups, but the change was not dose-related. Liver weight was similar in all groups, and relative tissue weight increased only in high-exposure group. Liver histopathologic findings were similar in all dose groups, and not different from the control group.

Dudley et al. (2001) administered JP-8 to female B6C3F1 and DBA/2 mice by oral gavage at 1 or 2 g/kg per day JP-8 for 7 days. Oral JP-8 was associated with a significant increase in body weight in the 1- and 2-g/kg groups of B6C3F1 mice but did not markedly affect body weight gain in DBA/2 mice. Liver weights were not reported, but both doses of JP-8 increased relative liver weight in both strains of mice. Measurement of hepatic cytochrome P1A1 with Western blot analyses revealed no marked change in expression. Reported tissue and body weight changes were not dose-related, and the doses used and the route of administration are of questionable relevance to occupationally-exposed humans. Witzmann et al. (2000) exposed male Sprague-Dawley rats to JP-8 aerosol with a mass median aerodynamic diameter of 1.7-1.9 mm (M. Witten, University of Arizona, personal communication, 2002) by inhalation for 6 hr/day, 5 days/wk for 6 wk. The concentration of JP-8 in the chamber was 1,000 mg/m^3. Eighty-two days after exposure, there was no significant change in body weight, and the general health of the rats appeared normal. According to results of electrophoresis, protein mass "fingerprinting," and sequence tag analysis, hepatic lamin L83 abundance was significantly decreased and lamin L603 abundance was increased. However, total lamin A abundance was not markedly altered by JP-8. Only one measurement time (82 hr after exposure) and one concentration were studied. The relevance of the Witzmann et al. (2000) findings for human risk assessment is not known.

Several animal studies have examined the effect of kerosene, the primary substance in JP-8, on liver function. Reductions in blood glucose concentrations were noted in rats after subchronic inhalation of kerosene vapor (and possibly some aerosol) at a mean of 58 mg/m^3 (Starek and Vojtisek 1986 as cited in ATSDR 1998). Increased blood lactate and pyruvate concentrations were observed in rats exposed to kerosene at a mean of 231 mg/m^3, but not at a mean of 58 mg/m^3. The authors speculate that decreased circulating glucose concentrations were associated with increased glycolysis and the inhibition of gluconeogenesis. The effect of kerosene on glycolysis is supported by the findings of increased concentrations of lactate and pyruvate in the blood and liver and increased lactate dehydrogenase activity in the liver. The authors suggested that increased glycolysis was a result of inhibition of cellular respiration by kerosene. In another study, rats and dogs were exposed to deodorized kerosene at 100 mg/m^3 for 6 hr/day, 5 days/wk for 13 wk (Carpenter et al. 1976). No histopathologic changes were observed in the livers of the rats or dogs, and no liver weight changes were noted in the dogs.

EFFECTS OF IN VITRO EXPOSURE TO JP-8

Grant et al. (2000) examined the in vitro cytotoxic potential of JP-8 in an H4IIE liver cell line. The H4IIE cell line is an established model used to assess hepatic function and responds to polycyclic aromatic hydrocarbons. In 72-hr viability assays, the concentration of JP-8 producing 50% inhibition (IC_{50}) of growth in H4IIE cells was 12.6 ± 0.4 μg/mL. The relevance of the in vitro findings for humans is not known.

CONCLUSIONS AND RECOMMENDATIONS

In one experimental animal study, F344 rats and C57BL/6 mice continuously exposed to JP-8 vapors at concentrations up to 1,000 mg/m^3 for up to 90 days did not show significant changes in hepatic function or structure. In another study, liver weights in male F344 rats exposed to JP-8 aerosols at up to 1,000 mg/m^3 for 1 hr per day for 28 days were not significantly different from liver weights in control animals. There were no significant alterations in serum aspartate aminotransferase and alanine aminotransferase activities, indicators of hepatic function, and there were no marked changes in the liver histopathologic findings and cytochrome P450 content, a measure of xenobiotic metabolism. No liver toxicity was observed in rats and mice exposed to JP-4 vapors at up to 5,000 mg/m^3 for 6 hr/day, 5 days/wk for 12 mo.

The Subcommittee on Permissible Exposure Levels for Military Fuels, which wrote the 1996 National Research Council report *Permissible Exposure Levels for Selected Military Fuel Vapors*, used the latter study as a basis for derivation of the interim PEL. On the basis of a no-observed-adverse-effect level of 5,000 mg/m^3 identified in rats given JP-4 and applying an uncertainty factor of 10 for interspecies extrapolation, the PEL was 500 mg/m^3 (no intraspecies uncertainty factor was used).

The subcommittee recommends that liver toxicity be evaluated in experimental animals exposed to JP-8 vapors and mixtures of vapors and aerosols by the inhalation route. Because inhalation exposures greater than approximately 1,000 mg/m^3 for pure JP-8 vapors are difficult to achieve, the Air Force should consider conducting studies with saturated vapor atmospheres on larger numbers of animals or employ longer exposure durations (i.e., longer than 90 days) to increase the power of the studies for observing adverse responses in various organ systems.

REFERENCES

ATSDR (Agency for Toxic Substances and Disease Registry). 1998. Toxicological Profile for Jet Fuels (JP-5 and JP-8). U.S. Department of Health and Human Services, Public Health Service, Agency for Toxic Substances and Disease Registry, Atlanta, GA.

Bruner, R.H., E.R. Kinkead, T.P. O'Neill, C.D. Fleming, D.R. Mattie, C.A. Russell, and H.G. Wall. 1993. The toxicologic and oncogenic potential of JP-4 jet fuel vapors in rats and mice: 12-month intermittent inhalation exposures. Fundam. Appl. Toxicol. 20(1):97-110.

Butler, M.A., C.A. Flugel, E.F. Krieg, J.E. Snawder, and J.S. Kesner. 2001. Gene-environment interactions and exposure to JP8 jet fuel. Pp. 76-80 in JP8 Final Risk Assessment. The Institute of Environmental and Human Health (TIEHH), Lubbock, TX. August 2001.

Carpenter, C.P., D.L. Geary Jr., R.C. Myers, D.J. Nachreiner, L.J. Sullivan, and J.M. King. 1976. Petroleum hydrocarbon toxicity studies. XI. Animal and human response to vapors of deodorized kerosene. Toxicol. Appl. Pharmacol. 36(3):443-456.

Dosing, M., S. Loft, and E. Schroeder. 1985. Jet fuel and liver function. Scand. J. Work Environ. Health. 11(6):433-437.

Dudley, A.C., M.M. Peden-Adams, J. EuDaly, R.S. Pollenz, and D.E. Keil. 2001. An aryl hydrocarbon receptor independent mechanism of JP-8 jet fuel immunotoxicity in Ah-responsive and Ah-nonresponsive mice. Toxicol. Sci. 59(2):251-259.

Gaworski, C.L., J.D. MacEwen, E.H. Vernot, R.H. Bruner, and M.J. Cowan Jr. 1984. Comparison of the subchronic inhalation toxicity of petroleum and oil shale JP-5 jet fuels. Pp. 33-48 in Advances in Modern Environmental Toxicology, Vol. 6.

Applied Toxicology of Petroleum Hydrocarbons, H.N. MacFarland, C.E. Holdworth, J.A. MacGregor, R.W. Call, and M.L. Lane, eds. Princeton, NJ: Princeton Scientific Publishers.

Gibson, R.L., S. Shanklin, and R.L. Warner. 2001a. Health effects comparisons. Pp. 125-129 in JP-8 Final Risk Assessment Report. The Institute of Environmental and Human Health (TIEHH), Lubbock, TX. August 2001.

Gibson, R.L., S. Shanklin, and R.L. Warner. 2001b. Self-reported health status. Pp. 132-139 in JP-8 Final Risk Assessment Report. The Institute of Environmental and Human Health (TIEHH), Lubbock, TX. August 2001.

Grant, G.M., K.M. Shaffer, W.Y. Kao, D.A. Stenger, and J.J. Pancrazio. 2000. Investigation of in vitro toxicity of jet fuels JP-8 and jet A. Drug Chem. Toxicol. 23(1):279-291.

MacEwen, J.D., and E.H. Vernot. 1978. Toxic Hazards Research Unit Annual Technical Report. AMRL-TR-78-55. Aerospace Medical Research Laboratory, Wright-Patterson Air Force Base, Dayton, OH.

MacEwen, J.D., and E.H. Vernot. 1980. Toxic Hazards Research Unit Annual Technical Report. AMRL-TR-80-79. Aerospace Medical Research Laboratory, Wright-Patterson Air Force Base, Dayton, OH.

MacEwen, J.D., and E.H. Vernot. 1981. Toxic Hazards Research Unit Annual Technical Report. AMRL-TR-81-126. Aerospace Medical Research Laboratory, Wright-Patterson Air Force Base, Dayton, OH.

MacEwen, J.D., and E.H. Vernot. 1982. Toxic Hazards Research Unit Annual Technical Report. AMRL-TR-82-62. Aerospace Medical Research Laboratory, Wright-Patterson Air Force Base, Dayton, OH.

MacEwen, J.D., and E.H. Vernot. 1983. Toxic Hazards Research Unit Annual Technical Report. AMRL-TR-83-64. Aerospace Medical Research Laboratory, Wright-Patterson Air Force Base, Dayton, OH.

MacEwen, J.D., and E.H. Vernot. 1984. Toxic Hazards Research Unit Annual Technical Report. AMRL-TR-84-001. Aerospace Medical Research Laboratory, Wright-Patterson Air Force Base, Dayton, OH.

MacEwen, J.D., and E.H. Vernot. 1985. Toxic Hazards Research Unit Annual Technical Report. AMRL-TR-85-058. Aerospace Medical Research Laboratory, Wright-Patterson Air Force Base, Dayton, OH.

MacNaughton, M.G., and D.E. Uddin. 1984. Toxicology of mixed distillate and high-energy synthetic fuels. Pp. 121-132 in Advances in Modern Environmental Toxicology, Vol. 7. Renal Effects of Petroleum Hydrocarbons, M.A. Mehlman, G.P. Hemstreet III, J.J. Thorpe, and N.K. Weaver, eds. Princeton, NJ: Princeton Scientific Publishers.

Mattie, D.R., C.L. Alden, T.K. Newell, C.L. Gaworski, and C.D. Flemming. 1991. A 90-day continuous vapor inhalation toxicity study of JP-8 jet fuel followed by 20 or 21 months of recovery in Fischer 344 rats and C57BL/6 mice. Toxicol. Pathol. 19(2):77-87.

Mattie, D.R., G.B. Marit, C.D. Flemming, and J.R. Cooper. 1995. The effects of JP-8 jet fuel on male Sprague-Dawley rats after a 90-day exposure by oral gavage. Toxicol. Ind. Health 11(4):423-435.

NRC (National Research Council). 1996. Permissible Exposure Levels for Selected Military Fuel Vapors. Washington, DC: National Academy Press.

Parton, K.H. 1994. The Effects of JP-8 Jet Fuel Inhalation on Liver and Kidney Function in Male F-344 Rats. M.S. Thesis, University of Arizona. 76pp.

Snawder, J.E., and M.A. Butler. 2001. Sensitive early indicators of hepatic and kidney damage in workers exposed to jet fuel. Pp. 81-86 in JP-8 Final Risk Assessment Report. The Institute of Environmental and Human Health (TIEHH), Lubbock, TX. August 2001.

Starek, A., and M. Vojtisek. 1986. Effects of kerosene hydrocarbons on tissue metabolism in rats. Pol. J. Pharmacol. Pharm. 38(5-6):461-469.

Wall, H.G., A. Vingegar, and E.R. Kinkead. 1990. Evaluation of Toxic Effects in Rats and Mice Exposed to JP-4 Vapor for One Year. Toxic Hazards Research Unit Annual Technical Report. AMRL-TR-90-063. Aerospace Medical Research Laboratory, Wright-Patterson Air Force Base, Dayton, OH.

Witzmann, F.A., R.L. Carpenter, G.D. Ritchie, C.L. Wilson, A.F. Nordholm, and J. Rossi III. 2000. Toxicity of chemical mixtures: Proteomic analysis of persisting liver and kidney protein alterations induced by repeated exposure of rats to JP-8 jet fuel vapor. Electrophoresis 21(11):2138-2147.

8

Effects of Jet-Propulsion Fuel 8 on the Kidney

This chapter summarizes the findings on kidney toxicity of jet-propulsion fuel 8 (JP-8) and related fuels presented in the National Research Council report *Permissible Exposure Levels for Selected Military Fuel Vapors* (NRC 1996) and reviews additional studies, most of which were completed after the 1996 report was published. The subcommittee uses that information to assess the potential toxic effects of JP-8 on the kidney in humans.

SUMMARY OF STUDIES DISCUSSED IN THE 1996 NATIONAL RESEARCH COUNCIL REPORT

The National Research Council Subcommittee on Permissible Exposure Levels for Military Fuels reviewed studies on the toxic effects of the vapors of JP-5, JP-8, and diesel fuel marine (DFM) on the kidney (NRC 1996).

That subcommittee reported that data on potential adverse health effects of JP-8 on the kidney were sparse. No human studies had examined kidney toxicity of JP-8. An acute exposure to hydrocarbon-based solvents at high concentrations (doses not specified) has been reported in a case study to produce kidney failure (Beirne and Brennan 1972); the authors reported that a person exposed to jet fuel (type and dose not specified) while fueling aircraft in the U.S. Air Force had mild, reversible focal glomerulonephritis.

One animal study that examined potential adverse effects of JP-8 exposure on the kidney was identified. Mattie et al. (1991) showed that exposure to JP-8 causes kidney lesions in male rats. Male and female F344 rats and C57BL/6 mice were exposed to JP-8 vapor at 500 or 1,000 mg/m^3 for 90 days. After the 90-day exposure, a triad of lesions were found in the kidneys of male rats: dramatically exacerbated hyalin droplet formation, granular casts in the outer medulla, and increased incidence and severity of lesions undifferentiable from those of chronic progressive nephrosis. No such lesions were observed in female rats. In the male and female mice, no histopathologic lesions related directly to JP-8 were found. The increased incidence and severity of chronic progressive nephrosis persisted throughout the remainder of the lifetimes of the male rats. The kidney changes observed after 90 days were not reversible and were progressive. The severity of lesions was greater after the higher exposure. No kidney tumors were found in the study. No other animal studies of the effects of JP-8 on kidneys were identified in the 1996 report.

Several studies that examined kidney toxicity of jet fuels other than JP-8 were described in the 1996 report (Parker et al. 1981; Bogo et al. 1983; MacEwen and Vernot 1985; Bruner et al. 1993). The results of those studies were consistent with the results of the Mattie et al. (1991) study. Male rats exposed to JP-4, JP-5, or DFM vapors developed kidney lesions consistent with hyaline droplet degeneration and resembling what is known as alpha 2u-globulin nephropathy. The mechanisms that underlie the development of that lesion are believed to occur only to male rats. The 1996 subcommittee concluded that the lesion is not expected to occur in humans.

EFFECTS OF EXPOSURE TO JP-8 IN HUMANS

The effects of exposure to JP-8 on the human kidney were examined in a study recently completed by the U.S. Air Force. The preliminary results of that study are described below and summarized in Table 8-1.

Snawder and Butler (2001) collected venous blood and urine from 107 people who worked at six U.S. Air Force bases (AFBs): Davis Monthan AFB, Arizona; Seymour Johnson AFB, North Carolina; Langley AFB, Virginia; Pope, AFB, North Carolina; Little Rock AFB, Arkansas; and Hurlbert Field, Florida. The exposed workers were fuel tank-entry personnel with persistent exposure to jet fuel (defined as a 1-hr entry twice a week for at least 9 months). The unexposed group consisted of Air Force personnel who had no important occupational exposure to hydrocarbon solvents or fuels. The participants

TABLE 8-1 Effects of JP-8 Exposure on the Kidney in Humans[a]

Reference	Exposure Concentration	Exposure Duration	Results
Snawder and Butler 2001	Measurements taken in breathing zones of subjects; median concentration of naphthalene, 1.9 μg/m^3 (low-exposure group), 447 μg/m^3 (high-exposure group); median concentration of benzene, 3.1 μg/m^3 (low-exposure group), 242 μg/m^3 (high-exposure group)	High-exposure group had persistent exposure to JP-8 (defined as at least 1 hr twice a wk for 9 mo); low-exposure group had no significant exposure to jet fuel or solvents	Concentrations of urinary neph-alpha GST and pi-GST in subjects in normal range
Butler et al. 2001	Measurements taken in breathing zones of subjects; median concentration of naphthalene, 1.9 μg/m^3 (low-exposure group), 10.4 μg/m^3 (moderate-exposure group), 447 μg/m^3 (high-exposure group); median concentration of benzene: 3.1 μg/m^3 (low-exposure group), 7.45 μg/m^3 (moderate-exposure group), 242 μg/m^3 (high-exposure group)	High- and moderate-exposure groups had persistent exposure to JP-8; low-exposure group had no significant exposure to jet fuel or solvents	Analysis of CYP2EI, GSTT1, and NQO1 genotype data showed no statistically significant interaction between those genotypes, alpha-GST or pi-GST, and JP-8 exposure
Gibson et al. 2001a	Exposed group (5,706 people) had potential occupational exposure to JP-8; control group (5,706 people) did not work in occupations in which exposure to JP-8 would occur	Not reported	Analysis of medical records showed that subjects in all groups had similar health-care visit rates; no

(Continued)

TABLE 8-1 *Continued*

Reference	Exposure Concentration	Exposure Duration	Results
			differences among groups in kidney-related conditions
Gibson et al. 2001b	Measurements taken in breathing zones of subjects; median concentration of naphthalene, 1.9 $\mu g/m^3$ (low-exposure group), 10.4 $\mu g/m^3$ (moderate-exposure group), 447 $\mu g/m^3$ (high exposure group); median concentration of benzene, 3.1 $\mu g/m^3$ (low-exposure group), 7.45 $\mu g/m^3$ (moderate-exposure group), 242 $\mu g/m^3$ (high-exposure group)	High- and moderate-exposure groups had persistent exposure to JP-8; low-exposure group had no significant exposure to jet fuel or solvents	Analysis of self-assessment questionnaire did not report differences among groups in kidney-related conditions

[a]Data collected from volunteers (male and female active-duty Air Force personnel) at six Air Force bases in United States. Volunteers were divided into three exposure groups: high, moderate, and low. High-exposure group performed tasks associated with repairing aircraft fuel systems; moderate-exposure group performed tasks associated with fuel handling, distribution, recovery, and testing; low-exposure group does not routinely come into contact with jet fuel or solvents. Data were collected in morning before subjects went to work and after they completed their work for the day. Reported results are from preliminary analysis of data. Additional background information about these studies can be found in Appendix B.

Abbreviations: GST, glutathione-*S*-transferase; CYP2E1, cytochrome P 2E1; NQO1, NAD(P)H quinone oxidoreductase.

completed questionnaires providing job category, exposure level, and medical and demographic features. The exclusion criteria for participants were the presence of autoimmune disease, cancer, or diabetes and the use of immune-system altering drugs.

Blood and urinary samples were collected before and after the shift at the AFB and shipped to a National Institute for Occupational Safety and Health laboratory in Cincinnati, Ohio, for analysis. The markers of renal damage included urinary neph-alpha glutathione *S*-transferase (GST), an index of proximal epithelial cell function; pi-GST, an index of distal tubule epithelial cell function; and creatinine, a marker of renal function. With commercial immunoassay kits, the concentrations of urinary neph-alpha GST and pi-GST in control and exposed subjects were within the normal range. Butler et al. (2001) further categorized exposed workers into three groups. The high-exposure group comprised subjects routinely performing tasks associated with repair of aircraft fuel systems; the moderate-exposure group comprised subjects involved in fuel handling, distribution, recovery, and testing; and the low-exposure group does not normally come into contact with jet fuel. There was no statistically significant change in urinary alpha-GST or pi-GST concentrations among any of the groups. Genotype was not considered a factor in renal response to JP-8 exposure in humans. The creatinine concentrations in urine of exposed personnel after the shift were higher than the creatinine concentrations in urine of unexposed subjects. However, even the highest urinary creatinine concentrations were within the normal clinical range; they were often indicative of mild dehydration. Data indicated that those sensitive measures of risk did not detect any adverse effect of acute JP-8 exposure on human renal function.

Gibson et al. (2001a) examined the medical records of Air Force personnel occupationally exposed to JP-8 (5,706 people—242 women, 5,464 men) and compared them with records of unexposed personnel (5,706 people randomly selected from a population of 20,244 unexposed people). All were active-duty members of the Air Force. A preliminary assessment of data showed that the total number of medical visits and the types of specific diseases—including circulatory, cardiovascular, and urogenital—were not markedly different between groups.

Gibson et al. (2001b) also conducted a self-assessment questionnaire survey of 328 exposed people, categorized into high-, moderate-, and low-exposure groups (as described above). A preliminary assessment of data showed that, in both men and women, the incidences of genitourinary, circulatory, and cardiovascular illnesses were not different between the groups.

EFFECTS OF EXPOSURE TO JP-8 IN EXPERIMENTAL ANIMALS

In addition to the studies summarized above, experimental-animal studies have been conducted to examine the effects of JP-8 exposure on renal function; they are described below and summarized in Table 8-2.

Parton (1994) subjected male F344 rats to nose-only exposure to JP-8 aerosol (average particle size was 1.1054 ± 0.2918 mm) for 1 hr/day for 7 or 28 days at 500 mg/m^3 (low-dose group) or 1,000 mg/m^3 (high-dose group). Body weight gain in the 28-day groups was significantly reduced compared to controls, but the final body weights of the exposed groups were not markedly different from those of the concurrent control group. Kidney weights were not significantly different between exposed and control groups. However, the relative kidney weight was increased in the low- and high-dose groups exposed for 7 days and in the high-dose group exposed for 28 days. There was no significant change in renal weight for the low-dose group exposed for 28 days. Histologic examination showed that the changes in relative kidney weight were accompanied by an increase in hyalin droplet formation and in alpha 2u-globulin levels. There was no significant effect of JP-8 exposure on blood urea nitrogen measurements between exposed and concurrent control groups. The increase in relative kidney weight reflects the increase in alpha 2u-globulin, but renal function was not compromised. The hyalin droplet formation associated with alpha 2u-microglobulin is gender- and species-specific with no relevance to humans (Flamm and Lehman-McKeeman 1991).

Mattie et al. (1995) administered JP-8 to Sprague-Dawley rats by oral intubation at 750, 1,500, or 3,000 mg/kg/day for 90 consecutive days. Urine simples were collected 24 hr before sacrifice. Blood and tissue samples were obtained at sacrifice. With respect to renal function, serum sodium and chloride concentrations increased only in the high-dose group. Serum creatinine concentrations increased in the low- and middle-dose groups but not in the high-dose group; however, urinary creatinine and protein concentrations were not significantly altered by JP-8. Urinary pH was significantly lower in the middle- and high-dose groups. The absolute renal weights were not altered in the exposed groups, but a significant increase in kidney:body weight ratio was found in the middle- and high-dose groups. The increase in renal weight was associated histopathologically with the accumulation of hyalin droplets in the cytoplasm of epithelial cells in the proximal convoluted tubules. The renal-function test result changes were not dose-related, and the histopathologic

TABLE 8-2 Effects of Jet Fuel Exposure on the Kidney in Experimental Animals

Fuel Type	Species	Exposure Concentration	Exposure Duration	Effects	Reference
JP-8	Male and female F344 rats, male and female C57BL/6 mice	500 or 1,000 mg/m^3 (vapors, whole-body)	90 days continuously	Kidney lesions (hyalin droplets, granular casts in outer medulla, nephrosis) in male rats only; no kidney toxicity in female rats or male and female mice	Mattie et al. 1991
JP-8	Male F344 rats	500 or 1,000 mg/m^3 (aerosol, nose-only)	1 hr/day for 7 or 28 days	Body weight gain in rats exposed for 28 days significantly decreased; final body weights of exposed animals similar to those of control animals; relative kidney weight increased in animals exposed for 7 days and in animals exposed at high dose for 28 days; changes in relative kidney weight associated with increase in hyalin droplet formation and in alpha-2u-globulin; renal function not compromised	Parton 1994
JP-8	Male Sprague-Dawley rats	750, 1,500, or 3,000 mg/kg (by gavage)	90 days consecutively	Serum sodium and chloride concentrations increased in highest-dose group; serum creatinine concentrations increased in low- and middle-dose groups (but not in high-dose group); urinary creatinine and protein concentrations not significantly altered by exposure; urinary pH significantly lower in the middle- and high-dose groups; exposure did not alter absolute renal weights but produced	Mattie et al. 1995

(Continued)

TABLE 8-2 *Continued*

Fuel Type	Species	Exposure Concentration	Exposure Duration	Effects	Reference
				significant increase in kidney: body weight ratio in the middle- and high-dose groups; increased renal weight caused by accumulation of hyalin droplets	
JP-8	Male Swiss-Webster mice	1,000 mg/m^3 (aerosol, nose-only)	1 hr/day for 5 days	Exposure significantly altered abundance of 56 proteins; concentrations of 21 proteins increased, concentrations of 35 proteins decreased, compared with controls	Witzmann et al. 2000a
JP-8	Male Sprague-Dawley rats	1,000 mg/m^3 (vapor, whole-body)	6 hr/day, 5 days/wk for 6 wk	Renal GST homolog and 10-formyltetrahydrofate dehydrogenase increased in charge modification index; no change in abundance	Witzmann et al. 2000b
JP-8, JP-5	Male and female C3Hf/Bd mice	Undiluted or 50% diluted in cyclohexane (dermal)	3 applications/wk for 60 wk	Exposure produced significant increase in water consumption; animals treated with undiluted fuels had significantly increased number of kidney lesions	Easley et al. 1982
JP-5	Rats	24 mL/kg of body weight (by gavage)	1 dose	LD_{50} about 60 mL/kg or higher; cytoplasmic droplets occurred in kidneys of male rats; serum creatinine and blood urea nitrogen increased in male rats	Parker et al. 1981; Bogo et al. 1983

JP-5	Rats, mice	150 or 750 mg/m^3 (vapor)	90 days continuously	More than 75% of male rats exposed at either concentration showed nephrosis and tubular damage; no kidney toxicity in mice	MacEwen and Vernot 1985
JP-4	Male and female F344 rats, C57BL/6 mice	1,000 or 5,000 mg/m^3 (vapor)	6 hr/day, 5 day/wk for 12 mo	Alpha-2u-globulin nephropathy occurred in males rats only; no kidney toxicity in mice	Bruner et al. 1993

Abbreviations: GST, glutathione-*S*-transferase; LD_{50}, dose that is lethal to 50% of the test animals.

alterations are not relevant for human health risk assessment (Flamm and Lehman-McKeeman 1991).

Easley et al. (1982) dermally exposed male and female C3Hf/Bd mice to undiluted or diluted (50% weight/volume dilution in cyclohexane) JP-8 or JP-5 three times per week for 60 wk. After 30 wk, the mice exposed to JP-8 and JP-5 had significantly increased water consumption. After 60 wk, the animals treated with undiluted JP-8 and JP-5 had a significant increase in number of kidney lesions, compared with the control animals.

Witzmann et al. (2000a) exposed male Swiss-Webster mice to JP-8 aerosol with a median mass aerodynamic diameter (MMAD) of 1.7-1.9 µm (M. Witten, University of Arizona, personnel communication, 2002) by nose-only inhalation. The total daily exposure time was 1 hr for a total of 5 days at an average JP-8 concentration of 1,000 mg/m^3. Control mice were subjected to ambient air. By means of various techniques—including electrophoresis, protein digestion, matrix-assisted laser desorption and ionization mass spectrometric protein identification, and sequence tagging with electrospray mass spectrometry—JP-8 exposure was shown to alter the abundance of 56 proteins. The concentrations of 21 proteins increased, and the concentrations of 35 proteins decreased; these 56 proteins accounted for 6% of all total resolved proteins. Only a single time-point and only one concentration were used, and the relevance of the findings to human health risk is not established. In a later study, Witzmann et al. (2000b) exposed male Sprague-Dawley rats to JP-8 for 6 hr/day, 5 days/wk for 6 wk. The concentration of JP-8 in the exposure chamber was 1,000 mg/m^3. Eighty-two days after exposure, there was no change in body weights, and the general health of the rats appeared normal. By means of electrophoresis, protein mass "fingerprinting," and sequence tag analysis, renal GST homolog and 10-formyltetrahydrofate dehydrogenase showed an increase in charge modification index but not in abundance. Only a single measurement time (82 hr after exposure) and a single concentration of JP-8 were used. The relevance of these findings to human health risk is not established.

CONCLUSIONS AND RECOMMENDATIONS

F344 rats and C57BL/6 mice exposed on a continuous basis by inhalation to JP-8 vapors at concentrations of 500 or 1,000 mg/m^3 for 90 days showed induction of alpha 2u-globulin nephropathy in male rats but not in female rats or in male or female animals of other species. Alpha-2u-globulin-induced nephropathy occurs only in male rats and is not relevant to humans.

The subcommittee recommends that kidney toxicity be evaluated in experimental animals exposed to JP-8 vapors and mixtures of vapors and aerosols by the inhalation route.

REFERENCES

Beirne, G.J., and J.T. Brennan. 1972. Glomerulonephritis associated with hydrocarbon solvents. Arch. Environ. Health. 25(5):365-369.

Bogo, V., R.W. Young, T.A. Hill, C.L. Feser, J. Nold, G.A. Parker, and R.M. Cartledge. 1983. Pp. 46-66 in The Toxicity of Petroleum and Shale JP5. AFRRI SR83-26. Armed Forces Radiobiology Research Institute, Bethesda, MD.

Bruner, R.H., E.R. Kinkead, T.P. O'Neill, C.D. Fleming, D.R. Mattie, C.A. Russell, and H.G. Wall. 1993. The toxicologic and oncogenic potential of JP-4 jet fuel vapors in rats and mice: 12-month intermittent inhalation exposures. Fundam. Appl. Toxicol. 20(1):97-110.

Butler, M.A., C.A. Flugel, E.F. Krieg, J.E. Snawder, and J.S. Kesner. 2001. Gene-environment interactions and exposure to JP8 jet fuel. Pp. 76-80 in JP8 Final Risk Assessment. The Institute of Environmental and Human Health (TIEHH), Lubbock, TX. August 2001.

Easley, J.R., J.M. Holland, L.C. Gipson, and M.J. Whitaker. 1982. Renal toxicity of middle distillates of shale oil and petroleum in mice. Toxicol. Appl. Pharmacol. 65(1):84-91.

Flamm, W.G., and L.D. Lehman-McKeeman. 1991. The human relevance of the renal tumor-inducing potential of d-limonene in male rats: Implications for risk assessment. Regul. Toxicol. Pharmacol. 13(1):70-86.

Gibson, R.L., S. Shanklin, and R.L. Warner. 2001a. Health effects comparisons. Pp. 125-129 in JP-8 Final Risk Assessment Report. The Institute of Environmental and Human Health (TIEHH), Lubbock, TX. August 2001.

Gibson, R.L., S. Shanklin, and R.L. Warner. 2001b. Self-reported health status. Pp. 132-139 in JP-8 Final Risk Assessment Report. The Institute of Environmental and Human Health (TIEHH), Lubbock, TX. August 2001.

MacEwen, J.D., and E.H. Vernot. 1985. Investigation of the 1-hour emergency exposure limit of JP-5. Pp. 137-144 in Toxic Hazards Research Unit Annual Technical Report: 1985. AAMRL-TR-85-058. Aerospace Medical Research Laboratory, Wright-Patterson AFB, OH.

Mattie, D.R., C.L. Alden, T.K. Newell, C.L. Gaworski, and C.D. Flemming. 1991. A 90-day continuous vapor inhalation toxicity study of JP-8 jet fuel followed by 20 or 21 months of recovery in Fischer 344 rats and C57BL/6 mice. Toxicol. Pathol. 19(2):77-87.

Mattie, D.R., G.B. Marit, C.D. Flemming, and J.R. Cooper. 1995. The effects of JP-8 jet fuel on male Sprague-Dawley rats after a 90-day exposure by oral gavage. Toxicol. Ind. Health 11(4):423-435.

NRC (National Research Council). 1996. Permissible Exposure Levels for Selected Military Fuel Vapors. Washington, DC: National Academy Press.

Parker, G.A., V. Bogo, and R.W. Young. 1981. Acute toxicity of conventional versus shale-derived JP5 jet fuel: Light microscopic, hematologic, and serum chemistry studies. Toxicol. Appl. Pharmacol. 57(3):302-317.

Parton, K.H. 1994. The Effects of JP-8 Jet Fuel Inhalation on Liver and Kidney Function in Male F-344 Rats. M.S. Thesis, University of Arizona. 76pp.

Snawder, J.E., and M.A. Butler. 2001. Sensitive early indicators of hepatic and kidney damage in workers exposed to jet fuel. Pp. 81-86 in JP-8 Final Risk Assessment Report. The Institute of Environmental and Human Health (TIEHH), Lubbock, TX. August 2001.

Witzmann, F. A., M.D. Bauer, A.M. Fieno, R.A. Grant, T.W. Keough, M.P. Lacey, Y. Sun, M.L. Whiten, and R.S. Young. 2000a. Proteomic analysis of the renal effects of simulated occupational jet fuel exposure. Electrophoresis 21(5):976-984.

Witzmann, F.A., R.L. Carpenter, G.D. Ritchie, C.L. Wilson, A.F. Nordholm, and J. Rossi III. 2000b. Toxicity of chemical mixtures: Proteomic analysis of persisting liver and kidney protein alterations induced by repeated exposure of rats to JP-8 jet fuel vapor. Electrophoresis 21(11):2138-2147.

9

Effects of Jet-Propulsion Fuel 8 on Reproduction and Development

This chapter reviews studies on potential reproductive and developmental toxicity of jet-propulsion fuel (JP-8). The subcommittee uses that information to assess the toxic effects of JP-8 on human reproduction and development. The National Research Council's (NRC's) report *Permissible Exposure Levels for Selected Military Fuel Vapors* (NRC 1996) did not review reproductive and developmental effects of exposure to jet fuels. Another NRC report, *Evaluating Chemical and Other Agent Exposures for Reproductive and Developmental Toxicity* (NRC 2001), did review the potential reproductive and developmental toxicity of JP-8. In that report, a dosage that is unlikely to cause toxicity (only for effects that are observed at birth and only for short-term exposure) was calculated to be 1 mg/kg per day (equivalent to 0.8 ppm for humans, assuming 8-hr/day exposure, 100% absorption, 69-kg body weight, and respiratory minute volume of 0.42 mL/min per kilogram of body weight).

EFFECTS OF EXPOSURE TO JP-8 IN HUMANS

No studies were found that examined the potential for developmental toxicity or adverse reproductive effects of JP-8 or other jet fuels in women. One study has examined reproductive effects in 50 men exposed to jet fuel (type not specified): six sheet metal workers, six painters, 15 men who fuel jets, and 23 flight-line workers (LeMasters et al. 1999); eight unexposed men served as the control group. The men were exposed to a mixture of solvents and fuels. Breath samples from the men were collected and measured for total naphthas, benzene, 1,1,1-trichloroethane, methyl ethyl ketone, xylenes, toluene, and methylene chloride. All mean measures were less than 6 ppm. The following characteristics were analyzed: sperm production, structure, and function (sperm concentration, motility, viability, morphology, and morphometrics and stability of sperm chromatin). Measurements were made before exposure and after 15 and 30 wk of exposure. There was an increase in sperm concentration in the jet-fueler and flight-line groups and a decrease in sperm linearity in the jet-fueler group. The authors concluded that exposure to jet fuel did not have an apparent effect on semen quality for aircraft-maintenance personnel.

EFFECTS OF EXPOSURE TO JP-8 IN EXPERIMENTAL ANIMALS

Several studies have been conducted to assess the effects of JP-8 on the reproductive and developmental systems of experimental animals. They are described below and summarized in Table 9-1.

Reproductive Toxicity

Mattie et al. (2000) examined fertility in male and female rats exposed to JP-8. Male Sprague-Dawley rats were given JP-8 at 0, 750, 1,500, or 3,000 mg/kg daily by gavage for 70 days before mating and during mating with unexposed females (up to 90 days). After 90 days, the male rats were sacrificed, and sperm concentration, motile sperm concentration, percentage motility, velocity, linearity, maximal amplitude of lateral head displacement (ALH), mean ALH, and beat/cross frequency were measured. Other dimensions measured were mean radius, number of circular cells, percentage circular cells/motile cells, and percentage circular cells/all cells. Pregnancy rate and gestation duration were recorded for all mated females. There were no adverse clinical signs, except changes in body weight. Rats exposed at 750 mg/kg showed a significant decrease in body weights ($p < 0.05$). There were no differences between exposed groups and the control group in any of the sperm measures. Exposure to JP-8 at the concentrations administered to the male mating partners did not have an effect on fertility of unexposed females.

Female Sprague-Dawley rats were given JP-8 at 0, 325, 750, or 1,500 mg/kg daily by gavage for 21 wk (90 days before cohabitation and during

TABLE 9-1 Effects of Jet Fuel Exposure on Reproduction and Development in Experimental Animals

Fuel Type	Species	Exposure Concentration	Exposure Duration	Effects	Reference
REPRODUCTIVE EFFECTS					
JP-8	Male, female Sprague-Dawley rats	Males, 750, 1,500, 3,000 mg/kg per day; females, 325, 750, 1,500 mg/kg per day (gavage)	Males, 70 days before mating and during mating (up to 90 days); females, 90 days before mating and during mating, gestation, delivery, lactation	Males: no differences between exposed and control groups in sperm concentration, motile sperm concentration, percentage motility, velocity, linearity, maximal ALH, mean ALH, beat/cross frequency, mean radius, number of circular cells, percentage circular cells/motile cells, and percentage circular cells/all cells; no effect on fertility of unexposed female mating partners Females: no significant differences between exposed and control groups in pregnancy rates, gestation lengths, number of pups/litter, litter size, viability and survival of pups; pups from dams exposed at 1,500 mg/kg per day had significantly reduced body weight compared with controls	Mattie et al. 1995, 2000
JP-8	Male rats	1,000 mg/m^3	6 wk	No significant differences between exposed and control groups on sperm maturation, total counts, viability, and morphology	Briggs et al. 1999 (meeting abstract)

(Continued)

TABLE 9-1 *Continued*

Fuel Type	Species	Exposure Concentration	Exposure Duration	Effects	Reference
JP-8	Male rats	250, 500, 1,000 mg/m^3	6 hr/day, 7 days/wk for 90 days	No significant differences between exposed and control groups on sperm count and concentration; no pathologic findings in testes of treated animals; significant difference between the treated and control animals in sperm motility	Price et al. 2001 (meeting abstract)
JP-5	B6C3F1 mice	2,000-8,000 mg/kg (dermal)	5 applications/wk for 13 wk	No histologic changes in reproductive systems	NTP 1986, as cited in ATSDR 1998
HDS Kerosene	Male, female Sprague-Dawley rats	165, 330, 494 mg/kg (dermal)	Males, 8 wk starting 14 days before mating; females, 7 wk starting 14 days before mating, sacrificed on days 4-6 of lactation	No treatment-related effect on fertility; no treatment-related microscopic changes in testes or epididymides of adult male rats or in ovaries of adult female rats	Schreiner et al. 1997
DEVELOPMENTAL EFFECTS					
JP-8	Female Sprague-Dawley rats	500, 1,000, 1,500, 2,000 mg/kg per day (oral)	Days 6-15 of pregnancy	Maternal and fetal body weights were markedly reduced in 1,000-, 1,500-, and 2,000-mg/kg per day groups; number and type of fetal malformations and variations did not differ	Cooper and Mattie 1996

				significantly between groups; progressive increase in overall incidence of fetal alterations with increasing dose between 500- and 1,500-mg/kg per day groups, but not for 2,000-mg/kg per day groups	
Jet Fuel A	Rats	100, 400 ppm (vapor)	6 hr/day on days 6-15 of pregnancy	No embryonic, fetotoxic, or teratogenic effects observed	Beliles and Mecler 1982, as cited in Koschier 1999
HDS Kerosene	Male, female Sprague-Dawley rats	165, 330, 494 mg/kg (dermal)	Males, 8 wk starting 14 days before mating; females, 7 wk starting 14 days before mating, sacrificed on days 4-6 of lactation	No treatment-related developmental toxicity	Schreiner et al. 1997
Kerosene	Rats	0.76, 2.6 mg/L (106, 365 ppm)	6 hr/day on days 6-15 of pregnancy	No adverse developmental effects in dams or progeny	API 1979, as cited in Koschier 1999

Abbreviations: ALH, amplitude of lateral head displacement; HDS, hydrodesulfurized cohabitation, gestation, delivery, and lactation) (Mattie et al. 2000). Male mating partners were not exposed to JP-8. Pregnancy rate, gestation duration, size of litter, number of pups born dead, and pup weight were recorded. A number of hematologic and clinical chemistry characteristics were measured. Urine was collected and analyzed for pH, specific gravity, total protein, and creatine. A pathologic (clinical pathologic and histopathologic) examination was conducted on the females. There were no adverse clinical signs, except changes in body weight, and no mortality. Body weights decreased in female rats exposed at 1,500 mg/kg. There were no significant differences in pregnancy rates, gestation lengths, number of pups per litter, litter size, and viability and survival of pups between exposed groups and the control group. The pups from the dams exposed at 1,500 mg/kg had significantly lower body weight (10% lower) than the control group.

In another study, male and female Sprague-Dawley rats were given hydrodesulfurized (HDS) kerosene at 0, 165, 330, or 494 mg/kg daily by the dermal route (Schreiner et al. 1997). Males were treated for about 8 wk starting 14 days before mating. Females were treated for about 7 wk, also starting 14 days before mating; they were sacrificed on days 4-6 of lactation. Pathologic examinations of the reproductive organs from males and females were conducted. Pregnancy rate, gestation duration, size of litter, number of pups born dead, and pup weight were recorded. No clinical signs of systemic toxicity were observed. Skin irritation increased in a dose-dependent manner. All groups had a fertility index of at least 80%. No treatment-related microscopic changes were observed in the testes or epididymides of adult male rats or in the ovaries of adult female rats.

Briggs et al. (1999) exposed male rats to JP-8 by inhalation at 1,000 mg/m^3 for 6 wk. Computer-assisted sperm analysis (CASA) was used to analyze sperm quality end points. The authors concluded that exposure to JP-8 had no significant influences on sperm maturation, total counts, viability, motility, and morphology. Price et al. (2001) exposed male rats to JP-8 by inhalation at 0, 250, 500, or 1,000 mg/m^3 for 6 hr/day, 7 days/wk for 90 days. Following cessation of exposure, rat semen was analyzed with CASA. There were no significant differences between the treated and control groups in sperm count and concentration. There were no pathologic findings in the testes. There was a significant difference between animals treated at the highest concentration and control animals in sperm motility.

Some information about reproductive effects of JP-8 can be obtained from toxicity studies that were not specifically designed to assess reproductive toxicity. Mattie et al. (1995) observed JP-8 had no effect on testis weight or histopathologic findings in a 90-day gavage study in male rats exposed at 0, 750, 1,500, or 3,000 mg/kg daily by gavage for 90 days. A study by the National Toxicology Program (NTP 1986 as cited in ATSDR 1998) found no histologic changes in the reproductive systems of mice treated dermally with JP-5 at 2,000-8,000 mg/kg, five times per week for 13 wk or at 250 or 500 mg/kg, five times per week for 103 wk.

Developmental Toxicity

Sprague-Dawley rats were given JP-8 orally at 0, 500, 1,000, 1,500, 2,000 mg/kg per day on days 6-15 of pregnancy (Cooper and Mattie 1996; NRC 2001). Dams exposed at 1,000 mg/kg per day or above gained significantly less body weight during pregnancy than did control rats. Several maternal deaths among exposed animals were attributed to the presence of JP-8 in the lungs. Fetal body weight at the two highest doses was significantly lower than in controls, but those doses were associated with even greater reduction in maternal weight gain during pregnancy. Fetal weight was reduced by 12% and 25%, and maternal gestational weight gain was reduced by 70% and 85% at 1,500 and 2,000 mg/kg per day, respectively. It is unclear whether the fetal weight reduction was causally associated with obvious signs of maternal toxicity. The numbers and types of congenital malformations and variations observed did not differ significantly between dose groups. A progressive increase in the overall incidence of fetal alterations (variations and malformations) with increasing dose was reported from 500 mg/kg per day to 1,500 mg/kg per day, but not at 2,000 mg/kg per day. It should be noted that the number of fetuses and litters exposed at 2,000 mg/kg per day and available for examination was much lower than in other dose groups because about one-third of the dams died; one surviving dam had a totally resorbed litter. Variations included dilated renal pelvis, ureter, and lateral ventricle; unossified sternebra; rudimentary 14th rib; fewer than four metatarsals; and external and subdural hematomas. Observed malformations included malformed sternum, missing centrum, hydronephrosis, ectopic heart, short tail, no tail, and encephalomyelcele.

Male and female Sprague-Dawley rats were exposed to HDS kerosene daily by the dermal route at 0, 165, 330, or 494 mg/kg diluted in mineral oil (Schreiner et al. 1997). Males were treated for about 8 wk, starting 14 days before mating. Females were treated for about 7 wk, also starting 14 days before mating, and were sacrificed on days 4-6 of lactation. There was no apparent developmental toxicity due to repeated topical application of HDS kerosene. There were no statistically significant differences between treated and control groups with regard to mean number of corpora lutea or number of implantation sites per dam. All groups had a live-birth index of at least 97%. Pups from control and treated groups had comparable birth weights and weight gains. There were no statistically significant differences between pups from treated and control groups in viability index or mean number of live pups per litter.

No signs of embryonic, fetotoxic, or teratogenic effects were observed in rats after exposure to jet fuel A vapor at 100 and 400 ppm on days 6-15 of gestation for 6 hr/day (Beliles and Mecler 1982 as cited in Koschier 1999).

Kerosene produced no adverse effects in dams or their progeny after inhalation exposure at 0.76 or 2.6 mg/L (106 or 365 ppm) for 6 hr/day on days 6-15 of gestation (API 1979 as cited in Koschier 1999).

CONCLUSIONS AND RECOMMENDATIONS

No studies were found in the literature that examined potential female reproductive effects or developmental effects of JP-8 or other jet fuels in humans. One study assessed male reproductive effects of inhalation of jet fuel (type not specified) and hydrocarbon solvents after 15 and 30 wk of exposure. In that study, exposure to jet fuel increased sperm concentration in workers who fueled jets and decreased sperm linearity in flight-line workers; exposure to jet fuels did not appear to affect semen quality in aircraft-maintenance workers.

Male and female Sprague-Dawley rats exposed to JP-8 by oral gavage at concentrations up to 1,500 (females) or 3,000 (males) mg/kg per day prior to and during mating and, in the case of the females, during gestation and lactation, showed a decrease in body weight, but no adverse effects on fertility were observed in either sex. Dermal exposure of rats to HDS kerosene at doses up to 494 mg/kg per day did not affect fertility in males or females exposed prior to and during mating and, in the case of the females, during gestation and lactation.

Maternal-gestational weight gain and fetal body weights were reduced in Sprague-Dawley rats exposed to JP-8 by oral gavage at 1,500 or 2,000 mg/kg per day on days 6-15 of pregnancy; the types of fetal abnormalities did not differ significantly between JP-8 dose groups and the unexposed animals, and there was a progressive increase in the overall incidence of abnormalities with increasing dose from 500 to 1,500 mg/kg per day, but not at 2,000 mg/kg per day. No developmental toxicity was reported in the offspring of Sprague-Dawley rats dermally exposed to HDS kerosene at doses up to 494 mg/kg per day. There are no developmental-toxicity studies that evaluate postnatal and long-term effects (such as neurologic effects) of in utero exposures.

Because of the paucity of data and because military personnel are occupationally exposed to JP-8, the subcommittee recommends that experimental-animal studies be conducted to determine the reproductive and developmental toxicity potential of JP-8.

REFERENCES

API (American Petroleum Institute). 1979. Teratology Study in Rats, Kerosine. Report No. 27-32175. Washington DC: American Petroleum Institute.

ATSDR (Agency for Toxic Substances and Disease Registry). 1998. Toxicological Profile for Jet Fuels (JP-5 and JP-8). U.S. Department of Health and Human Services, Public Health Service, Agency for Toxic Substances and Disease Registry, Atlanta, GA.

Beliles, R.P., and F.J. Mecler. 1982. Inhalation teratology of jet fuel A, fuel oil, and petroleum naphtha in rats. Pp. 233-238 in Proceedings of a Symposium-The Toxicology of Petroleum Hydrocarbons, H.N. MacFarland, C.E. Holdsworth, J.A. MacGregor, R.W. Call, and M.L. Lane, eds. Washington, DC: American Petroleum Institute.

Briggs, G.B., W.A. Price, A.F. Walsh, K.R. Still, and W.K. Alexander. 1999. Evaluation of JP-8 jet fuel potential to produce male reproductive toxicity using the computer-assisted sperm analysis system. Teratology 59(6):415.

Cooper, J.R., and D.R. Mattie. 1996. Developmental toxicity of JP-8 jet fuel in the rat. J. Appl. Toxicol. 16(3):197-200.

Koschier, F.J. 1999. Toxicity of middle distillates from dermal exposure. Drug Chem. Toxicol. 22(1):155-164.

Lemasters, G.K., D.M. Olsen, J.H. Yiin, J.E. Lockey, R. Shukla, S.G. Selevan, S.M. Schrader, G.P. Toth, D.P. Evenson, and G.B. Huszar. 1999. Male reproductive effects of solvent and fuel exposure during aircraft maintenance. Reprod. Toxicol. 13(3):155-166.

Mattie, D.R., G.B. Marit, C.D. Flemming, and J.R. Cooper. 1995. The effects of JP-8 jet fuel on male Sprague-Dawley rats after a 90-day exposure by oral gavage. Toxicol. Ind. Health 11(4):423-435.

Mattie, D.R., G.B. Marit, J.R. Cooper, T.R. Sterner, and C.D. Flemming. 2000. Reproductive Effects of JP-8 Jet Fuel on Male and Female Sprague-Dawley Rats After Exposure by Oral Gavage. AFRL-HE-WP-TR-2000-0067. Human Effectiveness Directorate, Air Force Research Laboratory, Wright Patterson AFB, OH. March.

NRC (National Research Council). 1996. Permissible Exposure Levels for Selected Military Fuel Vapors. Washington, DC: National Academy Press.

NRC (National Research Council). 2001. Evaluating Chemical and Other Agent Exposures for Reproductive and Developmental Toxicity. Washington, DC: National Academy Press.

NTP (National Toxicology Program). 1986. Toxicology and Carcinogenesis Studies of Marine Diesel Fuel and JP-5 Navy Fuel (CAS No. 8008-20-6) in B6C3F1 Mice (Dermal Studies). NTP 310. NIH 86-2566. Research Triangle Park, NC: National Toxicology Program/National Institutes of Health.

Price, W.A., G.B. Briggs, K.A. Grasman, and K.R. Still. 2001. Evaluation of reproductive toxicity from exposure of male rats to jet propulsion fuel JP-8 vapor. Toxicologist 60(1):251(1194).

Schreiner, C., Q. Bui, R. Breglia, D. Burnett, F. Koschier, P. Podhasky, L. Lapadula, R. White, M. Feuston, A. Kruegger, and S. Rodriquez. 1997. Toxicity evaluation of petroleum blending streams: Reproductive and developmental effects of hydrodesulfurized kerosene. J. Toxicol. Environ. Health 52(3):211-229.

10

Effects of Jet-Propulsion Fuel 8 on the Cardiovascular System

This chapter summarizes the findings on the cardiovascular system toxicity of jet-propulsion fuel 8 (JP-8) and related fuels presented in the National Research Council report *Permissible Exposure Levels for Selected Military Fuel Vapors* (NRC 1996) and reviews additional studies on cardiovascular system toxicity of JP-8 and related fuels. Those studies are summarized in Table 10-1. The subcommittee used the available information to assess the potential toxic effects of exposure to JP-8 on the cardiovascular system in humans.

SUMMARY OF STUDIES DISCUSSED IN THE 1996 NATIONAL RESEARCH COUNCIL REPORT

The National Research Council Subcommittee on Permissible Exposure Levels for Military Fuels reviewed studies on the toxic effects of hydrocarbon vapors on the cardiovascular system (NRC 1996; see Appendix A). Intentional or accidental inhalation of high concentrations of hydrocarbons has the potential to induce cardiac arrhythmias that can result in death. However, for the arrhythmias to occur, epinephrine must be released simultaneously with inhalation (Garb and Chenoweth 1948). No human studies that examined the car-

TABLE 10-1 Effects of Jet Fuel Exposure on the Cardiovascular System in Humans and Experimental Animals

Fuel Type	Species	Exposure Concentration	Exposure Duration	Effects	Reference
JP-8	Human, 5,706 exposed subjects and 5,706 unexposed subjects	Exposed group had potential occupational exposure to JP-8; control group did not work in occupations in which exposure to JP-8 would occur	Not reported	Medical records showed no increase in medical visits related to cardiovascular events	Gibson et al. 2001a[a]
JP-8	Human, 328 individuals	Measurements taken in breathing zones of subjects; median concentration of naphthalene, 1.9 $\mu g/m^3$ (low-exposure group), 10.4 $\mu g/m^3$ (moderate-exposure group), 447 $\mu g/m^3$ (high-exposure group); median concentration of benzene, 3.1 $\mu g/m^3$ (low-exposure group), 7.45 $\mu g/m^3$ (moderate-exposure group), 242 $\mu g/m^3$ (high-exposure group)	High- and moderate- exposure groups had persistent exposure to JP-8 (defined as at least 1 hr twice per wk for at least 9 mo); low-exposure group had no significant exposure to jet fuel or solvents	Data collected from self-assessment questionnaire; subjects in moderate- and high-exposure groups reported more heart palpitations and chest tightness than subjects in low-exposure group; odds ratios for subjects in moderate-exposure group, but not high-exposure group, were significantly greater than for low-exposure group	Gibson et al. 2001b[a]

Swedish military equivalent of JP-4	Human, 30 exposed subjects and 30 unexposed subjects	Average concentration, about 300 mg/m^3	Mean, 17 yr	Exposure may be associated with acute symptoms, such as palpitations and feeling of pressure in chest	Knave et al. 1978
JP-8	F344 rat, C57BL/6 mouse	500 and 1,000 mg/m^3 (via inhalation)	90 days continuously	No histopathologic changes to cardiovascuar system were observed	Mattie et al. 1991
JP-5	Sprague-Dawley rat	24 mL/kg (via oral gavage)	3 days	No increase in serum creatinine phosphokinase concentrations was observed	Parker et al. 1981
Deodorized kerosene vapor	Beagle, rat	20, 48, and 100 mg/m^3 (via inhalation)	6 hr/day, 5 days/wk for up to 67 day	No treatment-related changes in clinical pathologic and histopathologicl measures of cardiovascular system were observed; no electrocardiographic changes related to treatment were observed in dogs	Carpenter et al. 1976

[a]Background information about these studies can be found in Appendix B.

diovascular consequences of jet fuels were identified. Two studies in experimental animals examined the cardiovascular effects associated with oral administration of jet-propulsion fuel 5 (Parker et al. 1981). In one study, rats given JP-5 died from cardiovascular collapse unrelated to myocardial necrosis; in the other, rats given JP-5 at 24 mL/kg by oral gavage did not develop an increase in serum creatinine phosphokinase. Serum creatinine phosphokinase is indicative of cardiac-muscle damage. The Subcommittee on Permissible Exposure Levels for Military Fuels concluded that the animal data are not useful for determining permissible exposure levels (PELs), because the oral route of exposure was not directly relevant and the chemical composition of the liquid JP-5 differs from that of the vapors (NRC 1996).

EFFECTS OF EXPOSURE TO JET FUELS IN HUMANS

No studies of humans that directly evaluate the potential effects of jet-fuel exposure on the cardiovascular system have been conducted. Gibson et al. (2001a) examined the medical records of Air Force personnel occupationally exposed to JP-8 and compared them with records of unexposed (control) populations. The exposed group consisted of 5,706 people (242 women and 5,464 men). The control group consisted of 5,706 subjects randomly selected (equal numbers of men and women) from a cohort of 20,244 Air Force unexposed personnel. A preliminary assessment of medical records showed no increase in medical visits related to cardiovascular events. The study is limited by many factors, including information on potential confounders, completeness of health-event recording, differences among people in availability of health care, consequences of taking sick leave for health-care visits, differences in health-care-seeking behavior, and differences in amount of self-care or sensitivity to symptoms of illness.

Gibson et al. (2001b) conducted a health survey with a self-assessment questionnaire on 328 Air Force personnel (276 men and 52 women). The subjects were categorized as having high exposure (performed duties associated with aircraft fuel systems), moderate exposure (may have come into contact with jet fuel in the course of their duties), or low exposure (did not normally come into contact with jet fuel or other solvents while performing their duties). A preliminary assessment of the data showed that the total number of medical visits and the number of visits for specific reasons, including palpitations and chest tightness, were higher in the high- and moderate-exposure groups than in the low-exposure group. Odds ratios for people in the moderate-exposure group, but not the high-exposure group, were significantly greater than for those in the low-exposure group. The study is limited by the self-reporting of symptoms, failure to control for subject bias, and the fact that no exposure empirical data were collected.

Knave et al. (1978) conducted a cross-sectional epidemiologic study, focusing on the nervous system end points, of 30 exposed and unexposed workers in a jet-motor factory in Sweden. The workers were said to have been exposed to the Swedish military equivalent of JP-4 for a mean period of 17 years. The authors reported that acute symptoms—including respiratory tract symptoms (undefined), palpitations, and a feeling of pressure in the chest—may have been associated with the exposures.

EFFECTS OF EXPOSURE TO JP-8 AND KEROSENE IN EXPERIMENTAL ANIMALS

Male and female F344 rats and C57BL/6 mice were continuously exposed to JP-8 vapor by inhalation at 500 and 1,000 mg/m^3 for 90 days (Mattie et al. 1991). Rats were sacrificed after 2 wk, 2 months (mo), or 90 days of exposure or 9 mo or 21 mo after termination of exposure. Blood was taken from rats at the 2-wk, 2-mo, and terminal sacrifices for evaluation of blood chemistry. The exposures were well characterized with regard to concentration and chemical composition; no biologically significant changes were identified in clinical chemistry analyses. Forty tissues per animal were examined for histopathologic changes; treatment-related histopathologic changes were limited to the kidney. There was no evidence that subchronic inhalation of JP-8 vapors at concentrations higher than the interim PEL of 350 mg/m^3 caused adverse effects on the rat cardiovascular system.

To determine the potential adverse health effects of repeated inhalation of deodorized kerosene vapor, groups of 25 male rats and four male beagles were exposed for 6 hr/day, 5 days/wk for up to 67 days at 20, 48, and 100 mg/m^3 (Carpenter et al. 1976). End points included histopathologic findings (lungs, kidneys, liver, heart, spleen, adrenals, thyroids, trachea, and esophagus), hematologic and serum chemistry findings, and electrocardiographic findings (dogs only; baseline and after exposure). No treatment-related changes in clinical pathologic measures were found in rats or dogs. No treatment-related histopathologic lesions were found in either species. No electrocardiographic changes attributable to kerosene inhalation were noted in dogs (Carpenter et al. 1976).

CONCLUSIONS AND RECOMMENDATIONS

A preliminary comparison of medical records of Air Force personnel occupationally exposed to JP-8 with records of unexposed (control) personnel showed no increase in medical visits related to cardiovascular events. However, preliminary results of a health survey of Air Force personnel that used a self-assessment questionnaire showed that the total number of medical visits and the number of visits for specific reasons, including palpitations and chest tightness, were higher among high- and moderate-exposure groups than in the low-exposure group. The reported effects were not dose-related; the moderate-exposure group showed greater incidence of adverse effects than the high-exposure group. Many potential, uncontrolled biases are associated with those investigations, and the lack of adequate exposure data makes interpretation of the results difficult. The subcommittee recommends that when exposure-assessment data become available, the cardiovascular effects data be reevaluated.

The work of Mattie et al. suggests that continuous exposure of rats and mice to JP-8 vapor at up to 1,000 mg/m^3 for 90 days had no effect on the rodent cardiovascular system. Rats and dogs repeatedly exposed to kerosene at concentrations up to 100 mg/m^3 did not show any treatment-related effects.

The subcommittee recommends that cardiovascular toxicity be evaluated in experimental animals exposed to JP-8 vapors and mixtures of vapors and aerosols by the inhalation route.

REFERENCES

Carpenter, C.P., D.L. Geary Jr., R.C. Myers, D.J. Nachreiner, L.J. Sullivan, and J.M. King. 1976. Petroleum hydrocarbon toxicity studies. XI. Animal and human response to vapors of deodorized kerosene. Toxicol. Appl. Pharmacol. 36(3):443-456.

Garb, S., and M.B. Chenoweth. 1948. Studies on hydrocarbon-epinephrine induced ventricular fibrillation. J. Pharmacol. Exp. Ther. 94:12-18.

Gibson, R.L., S. Shanklin, and R.L. Warner. 2001a. Health effects comparisons. Pp. 125-129 in JP-8 Final Risk Assessment Report. The Institute of Environmental and Human Health (TIEHH), Lubbock, TX. August 2001.

Gibson, R.L., S. Shanklin, and R.L. Warner. 2001b. Self-reported health status. Pp. 132-139 in JP-8 Final Risk Assessment Report. The Institute of Environmental and Human Health (TIEHH), Lubbock, TX. August 2001.

Knave, B., B.A. Olson, S. Elofsson, F. Gamberale, A. Isaksson, P. Mindus, H.E.

Persson, G. Struwe, A. Wennberg, and P. Westerholm. 1978. Long-term exposure to jet fuel. II. A cross-sectional epidemiologic investigation on occupationally exposed industrial workers with special reference to the nervous system. Scand. J. Work Environ. Health 4(1):19-45.

Mattie, D.R., C.L. Alden, T.K. Newell, C.L. Gaworski, ans C.D. Flemming. 1991. A 90-day continuous vapor inhalation toxicity study of JP-8 jet fuel followed by 20 or 21 months of recovery in Fischer 344 rats and C57BL/6 mice. Toxicol. Pathol. 19(2):77-87.

NRC (National Research Council). 1996. Permissible Exposure Levels for Selected Military Fuel Vapors. Washington, DC: National Academy Press.

Parker, G.A., V. Bogo, and R.W. Young. 1981. Acute toxicity of conventional versus shale-derived JP5 jet fuel: Light microscopic, hematologic, and serum chemistry studies. Toxicol. Appl. Pharmacol. 57(3):302-317.

11

Genotoxic Effects of Jet-Propulsion Fuel 8

This chapter summarizes the findings on genotoxicity of jet-propulsion fuel 8 (JP-8) presented in the National Research Council report *Permissible Exposure Levels for Selected Military Fuel Vapors* (NRC 1996) and reviews additional studies, some of which were completed after the 1996 report was published. The studies are summarized in Table 11-1. The subcommittee used the body of evidence to assess the genotoxicity of JP-8 in humans.

SUMMARY OF STUDIES DISCUSSED IN THE 1996 NATIONAL RESEARCH COUNCIL REPORT

The National Research Council Subcommittee on Permissible Exposure Levels for Military Fuels reviewed studies relevant to the evaluation of the genotoxicity of JP-5, JP-8, and diesel fuel marine. The review included data from in vitro and in vivo rodent genotoxicity testing of JP-4, JP-5, and JP-8 (NRC 1996). Among the studies discussed in the 1996 report, the battery of in vitro and in vivo assays used by Brusick and Matheson (1978a) in testing JP-8 is the most relevant to the present assessment of JP-8, although the studies of JP-4 and JP-5 are also of interest.

TABLE 11-1 Genotoxic Effects of JP-8 Fuel in Humans and Experimental Animals

Fuel Type	Species/ Cell Line	Exposure Concentration	Exposure Duration	Effects	Reference
Hydro-carbons, jet-fuel derivatives	Human (34 male airport workers, 11 unexposed controls)	Benzene, 0.10 ± 0.05 mg/m^3; toluene, 0.13 ± 0.01 mg/m^3; xylenes, 0.13±0.02 mg/m^3, measured at Barcelona airport	9.77 yr (mean)	No increases in SCE, MN, or ras p21 protein levels were observed in exposed workers; significant difference in mean comet length and in genetic-damage index observed between exposed and unexposed workers	Pitarque et al. 1999
JP-4, solvents	Human (58 aircraft-maintenance workers, 8 unexposed controls)	All means below 6 ppm, as measured with industrial-hygiene methods	At least 30 wk	Exposure well below threshold limit values; small but statistically significant increase in frequency of SCE occurred after 30 wk of exposure in sheet-metal workers and painters; MN frequency in sheet-metal workers initially showed statistically significant increase but had decreased by 30 wk	Lemasters et al. 1997, 1999
JP-8	*Salmonella* strains, mouse lymphoma cells, human diploid WI-38 cells	Microbial assay, 0.001-5.0 μl/plate; mouse lymphoma assay, 0.01-	Microbial assay, 48 hr; mouse lymphoma assay, 4 hr; unscheduled	JP-8 not mutagenic in Ames-type reverse-mutation assay in *Salmonella* strains in either presence or absence of metabolic activation with rat liver S9; JP-8 toxic to most *Salmonella* strains at above 1 μL/plate; no gene mutation in mouse cells in L5178Y thymidine kinase	Brusick and Matheson 1978a

(Continued)

TABLE 11-1 *Continued*

Fuel Type	Species/ Cell Line	Exposure Concentration	Exposure Duration	Effects	Reference
		0.16 µl/ml; unscheduled DNA synthesis assay, 0.1-5.0 µL/mL	DNA synthesis assay, 1.5 hr	mouse lymphoma cell assay; JP-8 produced a significant, moderate increase in unscheduled DNA synthesis in WI-38 cells	
JP-8	H4IIE rat hepatoma cells	1-20 µg/mL	4 hr	JP-8 induced dose-dependent increase in mean comet tail moments, indicative of DNA damage; comet tail lengths and DNA strand breaks accumulated in presence of DNA repair inhibitors and JP-8; neither cytotoxicity nor significant apoptosis induced by JP-8	Grant et al. 2001
Various middle distillates	*Salmonella* strains	100-10,000 µg/plate	Not reported	Jet fuel A, JP-4, JP-5, MD API81-07 not mutagenic in *Salmonella* reverse-mutation assays; MDFs showed no mutagenic activity in *Salmonella*; straight-run MDFs nonmutagenic or marginally mutagenic; lightly refined paraffinic oil and C10-C14 normal paraffins negative in *Salmonella* at up to 10,000 µg/plate	IARC 1989; Brusick and Matheson 1978b; Pennzoil 1988; Nessel 1999; Deininger et al. 1991; McKee et al. 1994; McKee et al. 1989;

					Nessel et al. 1999
Various middle distillates	Mouse cells (L5178Y thymidine kinase mouse lymphoma cell assay)	Not reported	Not reported	Jet A fuel induced gene mutation in mouse cells in presence of metabolic activation (mouse or rat liver S9); straight-run kerosene positive in mouse assay in presence of metabolic activation; JP-4 not mutagenic in mouse assay; MD API 81-07 not mutagenic in mouse assay, did not induce SCEs in Chinese hamster ovary cells	IARC 1989; Koschier 1999; Brusick and Matheson 1978b; API 1984; Skisak 1991
Jet fuel A, middle distillates	CD-1 mice, Sprague-Dawley rats	Some 1.0-5.0 g/kg, some not reported	Some 24-27 hr, some not reported	Inhalation exposure of jet fuel A induced chromosomal aberrations in bone marrow of rats; exposure to turbo fuel A and C10-C14 normal paraffins by gavage did not induce MNs in CD-1 mouse bone marrow test; exposure of hydrodesulfurized kerosene by gavage induced chromosomal aberrations in bone marrow of mice; MD API 81-07 did not induce chromosomal aberrations in rat bone marrow but induced SCEs in $B6C3F_1$ mice	Conaway et al. 1984; IARC 1989; Koschier 1999; McKee et al. 1994; Nessel et al. 1999; Skisak 1991

Abbreviations: MN, micronucleus; SCE, sister chromatid exchange; MDF, middle distillate fraction.

Brusick and Matheson (1978a) observed that JP-8 was not mutagenic in the Ames-type reverse-mutation assay in *Salmonella typhimurium* strains TA1535, TA1537, TA1538, TA98, and TA100 in the presence or absence of metabolic activation with rat liver S9. JP-8 was toxic to most of the *Salmonella* strains at concentrations above 1 μL/plate. JP-8 was not mutagenic when tested in a yeast forward-mutation assay with *Saccharomyces cerevisiase* strain D4. It did not induce gene mutation in mouse cells in the L5178Y thymidine kinase mouse lymphoma-cell assay in the presence or absence of metabolic activation with mouse liver S9; it was moderately toxic at 0.16 μL/mL in this assay system. When tested for ability to induce unscheduled DNA synthesis (UDS) in WI-38 cells, a human diploid cell line, JP-8 produced a significant moderate increase in UDS, as measured by the incorporation of ^{3}H-thymidine, in either the presence or absence of mouse liver S9. The induction of UDS plateaued and was not dose-related; JP-8 toxicity was observed at 5 μL/mL. Brusick and Matheson interpreted the findings in WI-38 cells as suggesting that nonspecific DNA lesions were produced by JP-8. They tested JP-8 with the dominant-lethal test in rats and mice and reported that JP-8 was only moderately toxic at the doses tested and was negative in both mice and rats.

JP-4 was tested in the same battery of tests as those described for JP-8, with very similar results (positive solely in the WI-38-cell UDS assay) (Brusick and Matheson 1978b). JP-5 was not mutagenic in the Ames-type reverse-mutation assay in *Salmonella typhimurium* strains TA1535, TA1537, TA97, TA98, and TA100 in the presence or absence of metabolic activation with rat or hamster liver S9 (NTP 1986).

GENOTOXICITY IN HUMANS

In a study of 34 male workers exposed to hydrocarbons and jet-fuel derivatives at low concentrations at the Barcelona airport and 11 unexposed controls, Pitarque et al. (1999) measured *ras* p21 plasma protein concentrations, sister chromatid exchanges (SCEs), micronuclei (MNs), and DNA strand breaks, as detected by the Comet assay, in peripheral blood lymphocytes. No increases in *ras* p21 protein, SCEs, or MNs were observed in workers compared with controls. The frequency of binucleated cells with MNs was decreased in workers compared with controls. Statistically significant differences in mean Comet length and in the genetic damage index were observed between workers and controls. Confounding factors, such as age and smoking status, may have contributed to the findings; the mean age of the workers was 47.91 ± 4.22 (SEM) years (yr) compared with 34.87 ± 1.11 yr in controls, and the percentage of workers that smoked was 56.4% compared with 37.5% of the controls.

In a study of aircraft maintenance workers at a U.S. Air Force base exposed to solvents (Lemasters et al. 1997,1999) and JP-4 (Lemasters et al. 1997), 58 new hires were assessed for peripheral blood lymphocyte SCE and MN frequencies before starting work and again after 15 and 30 wk of work. Occupational exposures of the workers were well below the threshold limit values, with total solvent exposures all at less than 6 ppm, as measured by industrial-hygiene air methods. A time-dependent increase in SCEs was observed for the solvent- and fuel-exposed group compared with the unexposed group ($n = 8$). Most of the increases occurred in the sheet-metal (fuel-cell) workers ($n = 6$; $p = 0.003$), who had a 20% increase in SCEs, and the paint-shop workers ($n = 6$; $p = 0.05$). These workers had higher concentrations of solvents and fuel in their breath than workers in jet-fueling operations ($n = 15$) and the flight-line crew ($n = 23$). MNs in the jet-fueling operations workers went down over time. The authors concluded that the observations of increased SCEs in the exposed group might be due to chance, inasmuch as the increases were within ranges reported in the general population (Lemasters et al. 1999).

GENOTOXICITY STUDIES IN BACTERIA, YEAST, AND MAMMALIAN CELLS IN VITRO

JP-8

No additional data were identified on the genotoxicity of JP-8 in bacteria or yeast, other than the studies of Brusick and Matheson (1978a) discussed in the 1996 National Research Council report.

The ability of JP-8 to induce DNA damage in cultured mammalian cells has been investigated with the Comet (single-cell gel electrophoresis) assay. Grant et al. (2001) tested JP-8 in H4IIE rat hepatoma cells, which are capable of expressing many of the metabolic enzymes, including cytochrome P450-dependent oxidases, normally expressed in liver in vivo. JP-8, solubilized in ethanol at 0.1% (v/v), was applied to the H4IIE cells at 0-20 μg/mL for 4 hr, after which DNA damage was assessed with the Comet assay. JP-8 induced a dose-dependent increase in mean Comet tail moments in H4IIE cells; this indicates DNA damage. The authors reported that comet tail lengths increased and DNA strand breaks accumulated in the presence of DNA-repair inhibitors and JP-8 and concluded that JP-8 induces DNA damage, which can be mitigated by DNA repair. Neither cytotoxicity nor significant apoptosis was induced by JP-8.

Related Mixtures

Other jet fuels and related middle distillate fractions (MDFs) have been tested in bacteria and in vitro mammalian cell assays. As summarized by IARC (1989), jet fuel A and JP-5 were not mutagenic in *Salmonella* reverse-mutation assays. Neither was MD API 81-07, a hydrodesulfurized kerosene sample (Pennzoil 1988, as reviewed by Skisak 1991).

Nessel (1999) reviewed the data on several MDFs and found that they showed little or no mutagenic activity in *Salmonella*. Straight-run MDFs were either nonmutagenic (straight-run kerosenes: CONCAWE 1991, as cited by Nessel 1999) or marginally mutagenic (Deininger et al. 1991, as cited by Nessel 1999). McKee et al. (1994) evaluated five middle distillate materials, including turbo fuel A, in *Salmonella* strain TA98 in the presence or absence of hamster liver S9 and found that straight-run distillates were nonmutagenic; that is, they induced less than a doubling of revertant colonies. Lightly refined paraffinic oil (McKee et al. 1989) and C10-C14 normal paraffins (Nessel et al. 1999) were negative when tested in *Salmonella* at up to 10,000 μg/plate.

Jet fuel A induced gene mutation in mouse cells in the L5178Y thymidine kinase mouse lymphoma-cell assay in the presence but not in the absence of metabolic activation (mouse or rat liver S9), as reviewed by IARC (1989). Straight-run kerosene has also tested positive in the mouse lymphoma assay in the presence of metabolic activation (as summarized by Koschier 1999). MD API 81-07, a hydrodesulfurized kerosene, was not mutagenic in the mouse lymphoma assay (API 1984, as reviewed by Skisak 1991), nor did it induce SCE in Chinese hamster ovary cells (API 1988a, as reviewed by Skisak 1991).

IN VIVO GENOTOXICITY STUDIES IN ANIMALS

JP-8

No data were identified on the in vivo genotoxicity of JP-8 in animals other than the studies of Brusick and Matheson (1978b) discussed in the 1996 National Research Council report.

Related Mixtures

Other jet fuels and related MDFs have been tested for genotoxicity in animals in vivo. Jet fuel A, administered by inhalation, induced chromosomal aberrations in the bone marrow of male and female Sprague-Dawley rats (Conaway et al. 1984, as reviewed by IARC 1989, Koschier 1999). McKee et al. (1994) evaluated five middle distillate materials, including turbo fuel A, administered by gavage, in the CD-1 mouse bone marrow micronucleus test. No increases in the frequency of MNs were observed for any of the test materials in assessments 24, 48, or 72 hr after treatment. The authors did not see any evidence of bone marrow depression (McKee et al. 1994). C10-C14 normal paraffins administered by gavage did not induce MNs in the CD-1 mouse bone marrow micronucleus test in assessments 24, 48, or 72 hr after treatment in ether male or female mice (Nessel et al. 1999). Koshier (1999) reported that hydrodesulfurized kerosene, administered to mice by gavage, induced chromosomal aberrations in the bone marrow. Administration (route not specified) of MD API 81-07, a hydrodesulfurized kerosene, induced SCE in $B6C3F_1$ mice (API 1988b, as reviewed by Skisak 1991), but did not induce chromosomal aberrations in rat bone marrow (API 1984, as reviewed by Skisak 1991).

CONCLUSIONS AND RECOMMENDATIONS

The available data on genotoxicity in human populations exposed to jet fuels come from two relatively small studies of people exposed to jet fuels and a number of other solvents—one among workers at the Barcelona airport (type of jet fuel not specified) and one among workers exposed to JP-4 at a U.S. Air Force base. Both studies found slight genotoxic effects associated with exposure, but interpretation of the findings in those studies is complicated by a variety of factors, including small number of subjects studied and the multiple chemical exposures experienced by them in addition to exposure to jet fuel. In the case of the Barcelona airport study, the finding of increased DNA damage in workers is confounded by the higher mean age of workers than unexposed controls, and the higher percentage of workers than of controls who were smokers. In the case of the U.S. Air Force base study, the significance of observations of time-dependent increases in SCEs in some subgroups of workers is uncertain, given the small numbers of subjects in whom the increases were observed and the small magnitude of the increases (all were within the range of population controls).

Available data on the genotoxicity of JP-8 in animals, cultured cells, and prokaryotes indicate that JP-8 does not induce dominant lethal mutations in Sprague-Dawley rats or CD-1 mice, or mutations in *Salmonella typhimurium, Saccharomyces cerevisiae*, or the mouse lymphoma assay system. In vitro JP-8 exposure has been shown to induce DNA damage in human and rat cell lines, namely, induction of UDS in human diploid cell line WI-38 and DNA damage in H4IIE rat hepatoma cells. No published data regarding the genetic toxicity of JP-8 in vivo were identified.

A larger database is available on the genotoxicity of other jet fuels and related MDFs. Other jet fuels—including jet fuel A, and JP-5, and several middle distillates—were not mutagenic in *Salmonella typhimurium* strains. Mixed results have been reported for the in vitro mouse lymphoma assay: some materials tested positive (such as jet fuel A and straight-run kerosenes) and others negative (such as hydrodesulfurized kerosene). Mixed in vivo genotoxicity findings have been reported for other jet fuels and MDFs: inhalation exposure to jet fuel A induced chromosomal aberrations in rat bone marrow; gavage administration of hydrodesulfurized kerosene induced chromosomal aberrations in mice; gavage administration of turbo fuel A, C10-C14 normal paraffins, and other MDFs did not induce chromosomal aberrations in mice; and MD API 81-07, a hydrodesulfurized kerosene, did not induce chromosomal aberrations in the bone marrow of rats but did induce SCEs in the bone marrow of mice.

The subcommittee concludes that the available data are insufficient to draw a conclusion regarding the genotoxicity of inhaled JP-8. JP-8 has been shown to induce DNA damage in cultured mammalian cells, and some related mixtures (such as jet fuel A and straight-run kerosene) but not others (such as JP-4 and MD API 81-07, a hydrodesulfurized kerosene) have been shown to induce mutations in cultured mouse lymphoma cells. Some related mixtures (jet fuel A in rats, hydrodesulfurized kerosene in mice, and another hydrodesulfurized kerosene, MD API 81-07, in mice) but not others (turbo fuel A, MDFs, and C10-C14 normal paraffins in mice and hydrodesulfurized kerosene MD API 81-07 in rats) have been shown to be clastogenic in vivo. Therefore, the subcommittee recommends that the Air Force conduct in vivo genotoxicity studies by the inhalation route in two animal species to determine whether JP-8 is mutagenic, clastogenic, or capable of inducing other types of DNA damage via inhalation.

REFERENCES

API (American Petroleum Institute). 1984. Mutagenicity Evaluation Studies in the Rat Bone Marrow Cytogenetic Assay, in the Mouse Lymphoma Forward Mutation Assay: Hydrodesulfurized Kerosene, API Sample 81-07. API Med. Res. Publ. 32-30240. Washington, DC: American Petroleum Institute, Medicine and Biological Science Dept.

API (American Petroleum Institute). 1988a. Sister Chromatid Exchange (SCE) Assay in Chinese Hamster Ovary (CHO) Cell with API 81-07: Hydrodesulfurized Kerosene. API Med. Res. Publ. 35-32482. Washington, DC: American Petroleum Institute, Health and Environmental Science Dept.

API (American Petroleum Institute). 1988b. In Vivo Sister Chromatid Exchange (SCE) Assay with API 81-07: Hydrodesulferized Kerosene. API Med. Res. Publ. 36-30043. Washington, DC: American Petroleum Institute, Health and Environmental Science Dept.

Brusick, D.J., and D.W. Matheson. 1978a. Mutagen and Oncogen Study on JP-8. AMRL-TR-78-20. Prepared by Litton Bionetics, Inc., Kensington, MD, for the Aerospace Medical Research Laboratory, Wright-Patterson Air Force Base, OH.

Brusick, D.J., and D.W. Matheson. 1978b. Mutagen and Oncogen Study on JP-4. AMRL-TR-78-24. Prepared by Litton Bionetics, Inc., Kensington, MD, for the Aerospace Medical Research Laboratory, Wright-Patterson Air Force Base, OH.

Conaway, C.C., C.A. Schreiner, and S.T. Cragg. 1984. Mutagenicity evaluation of petroleum hydrocarbons. Pp 89-197 in Advances in Modern Environmental Toxicology, Vol. 6. Applied Toxicology of Petroleum Hydrocarbons, H.N. MacFarland, C.E. Holdsworth, J.A. MacGregor, R.W. Call, and M.L. Lane, eds, Princeton, NJ: Princeton Scientific.

CONCAWE (The Oil Companies' European Organization for Environment, Health, Safety). 1991. Middle Distillates – A Review of the Results of a CONCAWE Programme of Short-Term Biological Studies. Report 91/51. CONCAWE, Brussels, Belgium.

Deininger, G., H. Jungen, and R.P. Wenzel-Hartung. 1991. Middle Distillates: Analytical Investigations of Mutagenicity Studies. Research Reports No. 412-1. DGMK, Hamburg, Germany (as cited in Nessel et al 1999).

Grant, G.M., S.M. Jackman, C.J. Kolanko, and D.A. Stenger. 2001. JP-8 jet fuel-induced DNA damage in H4IIE rat hepatoma cells. Mutat. Res. 490(1):67-75.

IARC (International Agency for Research on Cancer). 1989. Jet fuel. Pp. 203-264 in Occupational Exposures in Petroleum Refining, Crude Oil and Major Petroleum Fuels. IARC Monographs on the Evaluation of Carcinogenic Risks to Humans, Vol. 45. Lyon: International Agency for Research on Cancer, World Health Organization.

Koschier, F.J. 1999. Toxicity of middle distillates from dermal exposure. Drug Chem. Toxicol. 22(1):155-164.

Lemasters, G.K., G.K. Livingston, J.E. Lockey, D.M. Olsen, R. Shukla, G. New, S.G. Selevan, and J.H. Yiin. 1997. Genotoxic changes after low-level solvent and fuel exposure on aircraft maintenance personnel. Mutagenesis 12(4):237-243.

Lemasters, G.K., J.E. Lockey, D.M. Olsen, S.G. Selevan, M.W. Tabor, G.K. Livingston, and G.R. New. 1999. Comparison of internal dose measures of solvents in breath, blood, and urine and genotoxic changes in aircraft maintenance personnel. Drug Chem. Toxicol. 22(1):181-200.

McKee, R.H., M.A. Amoruso, J.J. Freeman, and R.T. Przygoda. 1994. Evaluation of the genetic toxicity of middle distillate fuels. Environ. Mol. Mutagen. 23(3):234-238.

McKee, R.H., R.T. Plutnick, and R.T. Przygoda. 1989. The carcinogenic initiating and promoting properties of a lightly refined paraffinic oil. Fundam. Appl. Toxicol. 12:748-756.

Nessel, C.S. 1999. A comprehensive evaluation of the carcinogenic potential of middle distillate fuels. Drug Chem. Toxicol. 22(1):165-180.

Nessel, C.S., J.J. Freeman, R.C. Forgash, and R.H. McKee. 1999. The role of dermal irritation in the skin tumor promoting activity of petroleum middle distillates. Toxicol. Sci. 49(1):48-55.

NRC (National Research Council). 1996. Permissible Exposure Levels for Selected Military Fuel Vapors. Washington, DC: National Academy Press.

NTP (National Toxicology Program). 1986. Toxicology and Carcinogenesis Studies of Marine Diesel Fuel and JP-5 Navy Fuel (CAS No. 8008-20-6) in B6C3F1 Mice (Dermal Studies). NTP 310. NIH 86-2566. Research Triangle Park, NC: National Toxicology Program/National Institutes of Health.

Pennzoil. 1988. Mutagenicity Test on API 81-07 in the Modified Salmonella Microsome Mutation Assay for Petroleum Samples. Prepared for Pennzoil by Hazelton Laboratories America, Kesington, MD.

Pitarque, M., A. Creus, R. Marcos, J.A. Hughes, and D. Anderson. 1999. Examination of various biomarkers measuring genotoxic endpoints from Barcelona airport personnel. Mutat. Res. 440(2):195-204.

Skisak, C. 1991. The role of chronic acanthosis and subacute inflammation in tumor promotion in CD-1 mice by petroleum middle distillates. Toxicol. Appl. Pharmacol. 109(3):399-411.

12

Carcinogenic Effects of Jet-Propulsion Fuel 8

This chapter summarizes the findings on carcinogenicity of jet-propulsion fuel 8 (JP-8) presented in the National Research Council report *Permissible Exposure Levels for Selected Military Fuel Vapors* (NRC 1996) and reviews additional studies on JP-8 and related mixtures, some of which were completed after the 1996 report was published. The studies are summarized in Table 12-1. Because the available data on JP-8 are sparse, the subcommittee also reviewed carcinogenicity and genotoxicity data on some individual components of JP-8 that are identified as major components (by weight percent) or as carcinogens. The subcommittee used the body of available information to assess the carcinogenic potential of JP-8 in humans.

SUMMARY OF STUDIES DISCUSSED IN THE 1996 NATIONAL RESEARCH COUNCIL REPORT

The National Research Council (NRC) Subcommittee on Permissible Exposure Levels for Military Fuels reviewed studies relevant to the evaluation of the carcinogenicity of JP-5, JP-8, and diesel fuel marine (DFM) (NRC 1996). The review included epidemiologic studies of exposures to jet fuels and other

TABLE 12-1 Carcinogenic Effects of Fuels in Humans and Experimental Animals

Fuel Type	Species or Cell Line	Exposure Concentration	Exposure Duration	Effects	Reference
Jet fuel	Human (historical prospective cohort study of 2,176 men in Swedish armed forces)	Not reported	Not reported	No evidence of association between exposure to jet fuel and lymphatic malignancies	Selden and Ahlborg 1991 as cited in ATSDR 1998
Jet fuel	Human (population-based case-referent study of cohort of 3,726 cancer patients)	Not reported	Not reported	Screening-level analyses suggested association between kerosene exposure and stomach cancer, but result was not confirmed by in-depth analyses; screening analyses indicated that subjects with prior exposure to jet fuel (n = 43) had OR of 2.1 for colon cancer (n = 7), 2.1 for rectal cancer (n = 4), and 2.5 for kidney cancer (n = 7); in-depth analyses indicated association between jet-fuel exposure and kidney cancer (OR = 3.4) for workers exposed at substantial level (n = 6); dose-response relation observed for jet-fuel exposure and increased risk of kidney cancer	Siemaitycki et al. 1987
Jet fuel	Human (population-based case-control study;	Not reported	Not reported	Indication of excess risk of renal-cell carcinoma among aircraft mechanics and others with workplace exposures to jet fuel	Parent et al. 2000

	142 male patients with renal cancer, 1,900 controls with other types of cancers, 533 population-based controls)				
JP-8	F344 rat, C57B/6 mouse	500, 1,000 mg/m^3 (inhalation)	90 days continuously, then non-exposure period until 24 mo of age	No treatment-related tumors were observed in rats or mice of either sex; male rats had treatment-related accumulation of hyaline droplets in proximal convoluted tubular epithelium of kidney consistent with male alpha-2u-globulin nephropathy, a condition is specific to male rats	Mattie et al. 1991
JP-8	F344 rat	750, 1,500, 3,000 mg/kg per day (oral gavage)	90 days	No treatment-related tumors observed in rats; alpha-2u-globulin nephropathy, a condition specific to male rats, was observed	Mattie et al. 1995
JP-5	F344 rat, C57Bl/6 mouse, Beagle dog	150, 750 mg/m^3 (inhalation)	90 days continuously	No increase in tumors was observed in JP-5-treated mice	Gaworski et al. 1984, 1985
JP-5	$B6C3F_1$ mouse	250 or 500 mg/kg (dermal)	5 times/wk, 103 wk (males), 90 wk (females)	No increase in skin tumors was observed in JP-5-treated mice	NTP 1986

(Continued)

TABLE 12-1 *Continued*

Fuel Type	Species or Cell Line	Exposure Concentration	Exposure Duration	Effects	Reference
JP-4	F344 rat	1,000, 5,000 mg/m^3 (inhalation)	6 hr/day, 5 days/wk for 12 mo, followed by a non-exposure period of 12 mo	Males in the high-dose group had statistically nonsignificant increase in renal cell tumors; authors concluded that increase in tumors was due to alpha-2u-globulin accumulation and associated nephropathy	Bruner et al. 1993
Jet fuel A, lightly refined paraffinic oil	C3H mouse	Jet fuel A, undiluted; middle distillates, undiluted or 25% or 50% dilution in mineral oil or toluene; chemicals applied dermally	Neat 2 times/wk for 2 yr or on a intermittent schedule	Skin tumors found in 44% of mice treated with jet fuel A 2 times/wk and no tumors in control animals; skin tumors found in 2% of mice treated with jet fuel A intermittently; skin tumors found in 8% of mice treated with neat paraffinic oil and no tumors in mice treated with diluted material	Freeman et al. 1993
Jet fuel A, JP-4	C3H/HeN mouse	25 mg/dose (dermal)	3 times/wk for up to 105 wk	Skin tumors found in 28% of mice treated with shale-derived jet fuel A, 26% of mice treated with petroleum-derived jet fuel A, and skin-tumor latency was 79 wk; skin tumors found in	Clark et al. 1988

				50% of mice treated with shale-derived JP-4, 26% of mice treated with petroleum-derived JP-4, and skin-tumor latency was 84 wk	
Several middle distillates, including kerosene	C3H, CD-1 mouse	Undiluted or 50% or 28.6% dilutions in mineral oil	2 times/wk (undiluted), 7 times/wk (28.6%), and 4 times/wk (50%) for 52 wk or 2 yr following initiation with DMBA	Mice exposed to undiluted materials had significant increases in skin tumors, with incidence of 23-57%; exposure to diluted materials did not lead to increases in numbers of skin tumors	Nessel et al. 1998; Nessel et al. 1999
Lightly refined paraffinic oil	CD-1 mouse	Undiluted material	6 times over a 2-wk period followed by promotion with TPA (initiation assay) or DMBA treatment followed by 28-wk exposure (promotion study)	Oil was not a tumor initiator; had weak tumor-promoting activity (17% skin-tumor response, compared with 0% in controls)	Mckee et al. 1989

(Continued)

TABLE 12-1 *Continued*

Fuel Type	Species or Cell Line	Exposure Concentration	Exposure Duration	Effects	Reference
MD API 81-07 (middle distillates)	CD-1 mouse	Undiluted material	2 times/wk for 25 wk after initiation	MD API 81-70 had tumor-promoting effects; induced tumors were squamous cell carcinoma and papilloma of skin; treatment with dexamethasone inhibited tumor promotion	Skisak 1991

Abbreviations: DMBA, dimethylbenzanthracene; TPA, 12-*O*-tetradecanoyl-phorbol-13-acetate.

petroleum-based mixtures, such as gasoline; however, no studies of JP-5, JP-8, or DFM were located. Among the studies discussed in the 1996 report, the historical prospective cohort study of men in the Swedish armed forces (Selden and Ahlborg 1991) and the population-based case-referent study of Siemiatycki et al. (1987) appear to be the most relevant to the present assessment of JP-8. Selden and Ahlborg (1991) reported that total cancer incidences were generally lower than expected in the cohort of 2,176 men in the Swedish armed forces, and they observed no associations between aircraft fuel and cancer at any site. The 1996 report noted as possible limitations of that study the short followup period (9-10 years [yr]) and selection bias. Siemiatycki et al. (1987) investigated possible associations between exposures to 12 petroleum-derived liquids, including jet fuel and kerosene, and cancer among 3,726 subjects in Montreal. Screening analyses suggested an association between kerosene exposure and stomach cancer, but it was not confirmed by more in-depth analyses. Screening analyses indicated that people with exposure to jet fuel (e.g., aircraft mechanics and repairmen) (n = 43) had odds ratios (ORs) of 2.1 (90% CI, 0.9-5.1) for colon cancer (n = 7), 2.1 (0.6-7.4) for rectal cancer (n = 4), and 2.5 (1.1-5.4) for kidney cancer (n = 7). More in-depth analyses indicated an association between jet fuel and kidney cancer with an OR of 3.4 (1.5-7.6) for workers exposed at a substantial level (n = 6). A dose-response relation was observed for jet-fuel exposure and increased risk of kidney cancer, and the authors judged the strength of the evidence of this association as moderate to strong. On the basis of in-depth analyses, the authors also found a nonsignificant excess of colorectal cancers associated with jet-fuel exposure and noted a report of a slight excess of colorectal cancer (22 observed, 18 expected) among aircraft mechanics in Washington state. The subcommittee concluded that the data did not provide a consistent body of evidence sufficiently robust to support the conclusion that exposure to military jet fuels carries an excess risk of cancer at any site.

The Subcommittee on Permissible Exposure Levels for Military Fuels also discussed animal studies of chronic inhalation exposures to JP-4 and unleaded-gasoline vapor; studies of subchronic inhalation exposures to JP-8, JP-5, and JP-4; and studies of dermal exposure (skin painting) to JP-5 and DFM (NRC 1996). No lifetime inhalation animal bioassays of JP-5, JP-8, or DFM were located. Among the studies discussed in the 1996 report, the subchronic inhalation studies of JP-8 (Mattie et al. 1991) and the studies of JP-5 (Gaworski et al. 1984, 1985; NTP 1986) and JP-4 (Bruner et al. 1993) are the most relevant to the present assessment. Those studies are discussed below with additional studies included in the current assessment.

CARCINOGENICITY STUDIES IN HUMANS

No long-term studies of the chronic health effects, including cancer, of JP-8 exposure have been conducted. With regard to epidemiologic studies of related jet fuels, one additional study published after the release of the 1996 NRC report was identified (Parent et al. 2000). Numerous studies of the carcinogenic potential of gasoline streams and related middle distillates have appeared in the open literature.

Parent et al. (2000) conducted further analyses of the occupational information collected in association with the population-based case-control study reported by Siemiatycki et al. (1987), examining the association between occupational exposures and renal-cell cancer. Some 142 male patients with renal-cell carcinoma, 1,900 controls with cancer at other sites, and 533 population-based controls were interviewed for occupational histories and data on potential confounders. Multivariate logistic-regression models based on population, cancer controls, or a pool of both groups were used to estimate ORs. With regard to aviation fuel, the authors reported indications of excess risks among aircraft mechanics (OR, 2.8; 95% CI,1.0-8.4) and among people employed in defense services for more than 10 yr (OR, 3.0; 95% CI,1.2-7.4). Excess risk of renal-cell cancer was associated with workplace exposures to jet fuel (OR, 3.5; 95% CI, 1.4-8.7) and aviation gasoline (OR, 3.5; 95% CI, 1.4-8.6). The latter analyses were adjusted for nonoccupational and occupational potential confounders. The authors noted that the high degree of correlation within the study population between exposures to jet fuel and aviation gasoline precluded assessment of the risks posed by each independently.

The subcommittee is aware of a suspected cancer cluster in Fallon, Nevada, and that exposure to JP-8, originating from a naval base located in that town, is under investigation as a possible cause of the cluster (exposures to other chemicals are being investigated as well). Since 1997, sixteen persons currently or previously living in Fallon have been diagnosed with acute lymphocytic leukemia (ALL), a type of childhood cancer. One case of ALL would be expected approximately every 5 yr in Churchill County, where Fallon is located, based on the size of the population (Nevada State Health Division 2002). No scientific studies were found that examined a potential relationship between ALL and JP-8 exposure; therefore, the subcommittee could not reach any conclusion concerning exposure to JP-8 and this suspected cancer cluster.

CARCINOGENICITY STUDIES IN ANIMALS

No data are available on long-term rodent carcinogenicity studies of exposure to JP-8 by any route. Easley et al. (1982) reported that JP-8, JP-5, and

DFM were administered to mice by the dermal route for 60 wk; however, the extremely limited reporting of skin-tumor findings in all treatment groups combined renders these studies uninformative with regard to carcinogenicity. Subchronic studies of 90-day inhalation exposures to JP-8, with observation for an additional 20-21 months (mo), have been conducted in rats and mice of both sexes (Mattie et al. 1991, discussed in NRC 1996); and a 90-day gavage study of JP-8 in male rats has been reported (Mattie et al. 1995).

Given the absence of data from carcinogenicity studies on JP-8, data on other middle-distillate fraction-derived mixtures with various degrees of similarity to JP-8 are discussed briefly below, including studies of 12-mo inhalation exposures to JP-4 in rats and mice with observation for an additional 12 mo (Bruner et al. 1993, discussed in NRC 1996), subchronic studies of 90-day inhalation exposures to JP-5 in male and female rats and in female mice, with observation for up to an additional 21 mo (Gaworski et al. 1984, 1985, discussed in NRC 1996), and several long-term mouse-skin-painting bioassays of jet fuel A (Clark et al. 1988; Freeman et al. 1993), JP-4 (Clark et al. 1988), JP-5 (NTP 1986, discussed in NRC 1996), MD API 81-07, a hydrodesulfurized kerosene (API 1988, as cited by Skisak 1991), straight-run kerosene (Nessel et al. 1998), and other middle distillate fractions (Freeman et al. 1993).

Inhalation-Exposure Studies

JP-8 has not been tested in lifetime rodent carcinogenicity bioassays by the inhalation route. It has been tested in rats and mice of both sexes in studies with 90-day exposures and then observation until the age of 24 mo (Mattie et al. 1991, discussed in NRC 1996). F344 rats and C57BL/6 mice were exposed continuously to JP-8 vapor at 0, 500, or 1,000 mg/m^3 for 90 days and then allowed to recover until the age of 24 mo. No treatment-related tumors were seen in rats or mice of either sex. Male rats exhibited treatment-related accumulation of hyaline droplets in the proximal convoluted tubular epithelium of the kidney, which was consistent with male alpha 2u-globulin nephropathy. The short duration of exposure to JP-8 in the studies severely limits their usefulness for purposes of carcinogenicity assessment.

Other jet fuels have not been tested in lifetime rodent carcinogenicity bioassays by the inhalation route, but 12-mo exposure studies of JP-4 and 90-day continuous-exposure studies of JP-5 discussed in the 1996 National Research Council report are briefly described here. Bruner et al. (1993) exposed groups of 100 F344 rats of each gender and 100 C57Bl/6 mice of each gender to JP-4 at 1,000 or 5,000 mg/m^3 for 6 hr/day, 5 days/wk for 12 mo; animals were allowed to live unexposed for an additional 12 mo. In rats, an increase in renal-cell tumors (three renal-cell adenomas, one carcinoma, and one sarcoma versus none in controls), which did not reach statistical significance, was observed in high-dose males. The authors attributed that increase in renal-cell tumors to alpha 2u-globulin accumulation and associated nephropathy. In mice, a statistically significant ($p < 0.05$) increase in hepatocellular adenomas was observed in high-dose females (two of 83, one of 79, and eight of 80 for control, low-dose, and high-dose groups, respectively), and a single hepatocellular carcinoma occurred in the high-dose group.

Gaworski et al. (1984, 1985) exposed groups of male and female F344 rats and female C57BL/6 mice to petroleum- or shale-derived JP-5 continuously at 150 or 750 mg/m^3 for 90 days. Some animals were killed immediately after cessation of exposure, and others were allowed to live unexposed for an additional 19 or 21 mo. No treatment-related tumors were observed in rats or mice.

Dermal-Exposure Studies

One report of carcinogenicity studies of JP-8 administered dermally to mice was identified in the published literature (Easley et al. 1982), but the extremely limited reporting of skin-tumor findings render the studies uninformative with regard to carcinogenicity. Briefly, groups of 15 C3H/fBd mice of each gender received dermal applications of JP-8, JP-5, or DFM 3 times/wk for 60 wk undiluted or as a 50% weight/volume dilution in cyclohexane; controls received cyclohexane. The entirety of the information provided in the published report on tumor occurrence consists of the statement that "skin tumors occurred in only 34 of the 360 test mice and in only 1 of 60 cyclohexane control mice."

Jet fuels other than JP-8 have been tested for carcinogenicity in animal studies by the dermal route. In addition to the skin-painting studies of JP-5 by the National Toxicology Program (NTP 1986) discussed in the 1996 NRC report, studies of jet fuel A (Clark et al. 1988; Freeman et al. 1993) and JP-4 (Clark et al. 1988) are described here.

Jet fuel A, derived from either shale or petroleum, produced skin tumors (squamous cell carcinoma and fibrosarcoma) in groups of 25 C3H/HeN mice of each gender treated 3 times/wk at 25 mg/dose for up to 105 wk (Clark et al. 1988). Twenty-eight percent of the shale-derived and 26% of the petroleum-derived jet fuel A-treated mice developed skin tumors; the observed skin tumor latency was 79 wk.

Freeman et al. (1993) tested jet fuel A in the C3H mouse skin-painting model, using two treatment protocols. In the first protocol, jet fuel A was applied neat twice a week to the skin of C3H mice for 2 yr. In the second protocol, jet fuel A was applied intermittently; treatment was suspended when marked signs of dermal irritation were noted in 20% of the animals. In the first treatment protocol, jet fuel A produced tumors in 44% of the treated mice—and marked skin irritation—compared with 0% tumors in untreated and mineral-oil controls. In the second protocol, only 2% of the treated animals (n = 1) developed skin tumors. In the second protocol, animals received a lower total dose of jet fuel A than those in the first protocol.

JP-4, derived from either shale or petroleum, produced skin tumors (squamous cell carcinomas and fibrosarcoma) in groups of 25 C3H/HeN mice of each gender treated 3 times/wk at 25 mg/dose for up to 105 wk (Clark et al. 1988). Fifty percent of the shale-derived and 26% of the petroleum-derived JP-4-treated mice developed skin tumors; the observed skin tumor latency was 84 wk.

JP-5 was applied to the skin of male and female B6C3F_1 mice at 250 or 500 mg/kg, 5 times/wk in studies conducted by the NTP (NTP 1986). Male mice were treated for 103 wk; female mice were sacrificed at 90 wk because of excessive irritation and ulceration at the site of application. No increase in skin tumors was observed in JP-5-treated mice, and the NTP concluded that "under the conditions of these 2-yr dermal studies, JP-5 navy fuel at doses of 250 and 500 mg/kg provided no evidence of carcinogenicity for male and female B6C3F1 mice."

As summarized by Nessel (1999), middle distillate fractions (MDFs) have been tested in numerous lifetime mouse skin-painting studies over the last 20 yr. Early mouse skin-painting studies documenting the carcinogenicity of MDFs in mouse skin include those of Lewis et al. (1984) and Biles et al. (1988), as cited by Nessel (1999). MD API 81-07, a hydrodesulfurized kerosene, was also shown to induce skin tumors in a C3H/HeJ mouse skin-painting 2-yr bioassay in 50% of the animals with a tumor latency of 76 wk (API 1988, as cited by Skisak 1991).

An MDF known as lightly refined paraffinic oil was tested in the C3H mouse skin-painting model, applied neat and in 25% and 50% dilutions in mineral oil or in toluene (Freeman et al. 1993). The neat lightly refined paraffinic oil induced tumors in 8% of the treated mice; no tumors were observed in animals that received the diluted material. Skin irritation was observed in animals that received either the neat material or material diluted in toluene but not in animals that received material diluted in mineral oil.

The role of skin irritation in the development of skin tumors was investigated by Nessel et al. (1998). In lifetime C3H mouse skin-painting studies, MDFs, including a straight-run kerosene, were applied neat and in 50% and 28.6% dilutions. Treatment with the neat straight-run kerosene induced skin tumors and skin irritation; treatment with the diluted material produced neither skin tumors nor irritation. In followup studies described by Nessel (1999), equal weekly doses of irritating (neat), minimally irritating (50% dilution), and nonirritating (28.6% dilution) MDFs, including a straight-run kerosene, were applied to the skin of C3H mice for 2 yr. Skin tumors were induced in mice that received the neat straight-run kerosene but not in mice that received equal doses of the straight-run kerosene in a diluted, nonirritating form.

Oral-Exposure Studies

JP-8 has not been tested in lifetime rodent carcinogenicity bioassays by the oral route. A 90-day gavage study in male Sprague-Dawley rats, although inadequate for purposes of carcinogenicity assessment, reported that JP-8 treatment was associated with the development of alpha 2u-globulin nephropathy (Mattie et al. 1995).

OTHER RELEVANT DATA

Other relevant data not included in the 1996 National Research Council report but considered in the present assessment include those from tumor-initiation and -promotion studies of jet fuels and other middle distillates and those on the carcinogenicity and genotoxicity of several individual components of JP-8.

Tumor Initiation and Promotion

JP-8 has not been tested for tumor-initiating or -promoting activity. Jet fuel A has been tested for tumor promoting activity in the CD-1 mouse model of skin tumors initiated by dimethylbenzanthracene (DMBA) (Nessel et al. 1999). Other MDFs have been tested for tumor-promoting activity in the same model system. In addition, hydrodesulfurized kerosene, hydrodesulfurized middle distillates, and lightly refined paraffinic oil have been tested for initiating activity with the model. Those studies are briefly described below.

Nessel et al. (1999) conducted a 1-yr tumor-promotion study of jet fuel A in CD-1 mice, comparing equal weekly doses of irritating and minimally irritating or nonirritating test material, to assess whether tumor promotion occurred as a secondary response to irritation. Jet fuel A was applied to DMBA-initiated CD-1 mouse skin at 100% 2 times/wk, or 50% dilution in mineral oil 4 times/wk, or 28.6% dilution in mineral oil 7 times/wk. Jet fuel A (100%) was very irritating to the skin and was an effective tumor promoter: about 40% of treated mice developed squamous cell carcinomas or papillomas. Diluted jet fuel A was not irritating to the skin, nor did it have any tumor-promoting effects.

As summarized by Nessel (1999), MDFs have been tested in several initiation-promotion mouse skin-painting studies designed to investigate the multistage process of tumorigenesis. Hydrodesulfurized kerosene (MD API 81-07) and hydrodesulfurized middle distillates (MD API 81-10) were tested for tumor-initiating and tumor-promoting activity in CD-1 mouse skin-painting studies (API 1989, as summarized by Nessel 1999). Both test materials were negative in the initiation assay compared with acetone-treated controls. Both hydrodesulfurized kerosene (MD API 81-07) and hydrodesulfurized middle distillates (MD API 81-10) were strong promoters compared with toluene-treated controls after initiation with DMBA.

McKee et al. (1989) tested the initiating and promoting activity of a lightly refined paraffinic oil in the CD-1 male mouse skin-painting model. In the initiator test, the lightly refined paraffinic oil was applied to mouse skin 6 times over a 2-wk period, and then the promoter 12-0-tetradecanoylphorbol 13-acetate (TPA) was administered for a period of 1 yr. Lightly refined paraffinic oil was not a tumor initiator in this assay: only three of 30 animals that were treated with the test material and TPA developed skin tumors compared with nine of 30 control animals that were treated with acetone and TPA. In the promoter test, the lightly refined paraffinic oil was applied to DMBA-treated mouse skin in a 28-wk study. The lightly refined paraffinic oil had weak promoting activity, producing a 17% skin tumor response (five of 30 mice) compared with a 0% skin tumor response (none of 30 mice) in DMBA-treated mice that did not receive promoter treatment ($p = 0.026$, one-tailed test) (McKee et al. 1989).

Lightly refined paraffinic oil and C10-C14 normal paraffins were tested in 1-yr tumor-promotion studies in DMBA-treated CD-1 mice (Nessel et al. 1999). Equal weekly doses of irritating and minimally irritating or nonirritating test materials were compared to assess whether tumor promotion was a secondary response to these effects. Test materials were applied to CD-1 mouse skin at 100% 2 times/wk at 50% dilution in mineral oil 4 times/wk, or at 28.6% dilution in mineral oil 7 times/wk. Both lightly refined paraffinic oil and C10-C14 normal paraffins were tumor promoters, and both were irritating to the skin when applied undiluted. Dilution greatly reduced skin irritation and tumor-promoting activity.

Skisak (1991) tested the tumor-promoting activity of MD API 81-07, a hydrosulfurized kerosene, in CD-1 mouse skin treated with DMBA. MD API 81-07 was applied 2 times/wk for 25 wk to the skin of treated mice. MD API 81-07 had tumor-promoting effects, inducing squamous cell carcinoma and papilloma of the skin. Acanthosis, a uniform thickening of the epidermis due to hyperplasia of the stratum spinosum, was the most common finding in MD API 81-07 treated animals. Inflammation of the skin was observed in some animals, as were excessive numbers of inflammatory cells in the dermis (mixed population of neutrophils, lymphoid cells, histocytes, and mast cells). The author noted that subacute inflammation at early to midstudy points did not correlate well with tumor incidence and concluded that subacute inflammation was not a significant factor in tumor promotion by MDFs such as MD API 81-07. The author suggested that induction of a lasting, mild hyperplasia is an essential but not sufficient requirement for development of skin tumors in this initiation-promotion model. Treatment with dexamethasone, a potent antimitotic and anti-inflammatory agent that inhibits mouse epidermal DNA synthesis, reduced acanthosis and completely inhibited tumor promotion by MD API 81-07.

Carcinogenicity and Genotoxicity of Individual Components of JP-8

The available data on the carcinogenicity of JP-8 are sparse. In light of that sparseness, and the small amount of data available on related mixtures, such as other jet fuels and MDFs, the carcinogenicity and genotoxicity of some individual components of JP-8 that are identified as being among the top 10 constituents of the liquid fuel (by weight percentage) or as carcinogens are briefly discussed below.

Benzene

Benzene is present at low concentrations in JP-8, generally at 0.1-0.8 wt %. The International Agency for Research on Cancer (IARC) has classified benzene as a known human carcinogen (Group 1) on the basis of sufficient evidence that benzene causes leukemia in humans and sufficient evidence of carcinogenicity in animals (IARC 1987).

Butylbenzene

Butylbenzene is one of the top 10 constituents of JP-8 (by weight percentage). No data on the carcinogenicity of butylbenzene were identified in the published literature. *Tert*-butylbenzene was not mutagenic when tested in five *Salmonella* strains and two strains of *Escherichia coli*. It did not induce mitotic gene conversion in *Saccharomyces cerevisiase*, and it did not induce chromosomal aberrations in rat liver (RL1) cells in vitro (HSDB 2001).

Decane

Decane is one of the top 10 constituents of JP-8 (by weight percentage). It exhibited cocarcinogenicity by enhancing the mouse skin carcinogenicity of benzo[a]pyrene (Van Duuren and Goldschmidt 1976) and of ultraviolet light (Bingham and Nord 1977). Decane was also tested as a tumor promoter in a two-stage carcinogenesis assay, and found to have tumor-promoting activity (Van Duuren and Goldschmidt 1976).

Dodecane

Dodecane is one of the top 10 constituents of JP-8 (by weight percentage). It exhibited cocarcinogenicity by enhancing the mouse skin carcinogenicity of benzo[a]pyrene (when used as the diluent) (Bingham and Falk 1969) and of ultraviolet light (Bingham and Nord 1977).

Ethylbenzene

Ethylbenzene is present at low concentrations in JP-8. IARC has classified it as a possible human carcinogen (Group 2B) on the basis of sufficient evidence of carcinogenicity in experimental animals and inadequate evidence in humans (IARC 2000). In 2-yr inhalation bioassays conducted by the NTP, increased incidences of renal tumors and testicular adenomas were observed in male rats exposed to ethylbenzene at 750 ppm, and the incidences of several tumor types in the lung, liver, thyroid, and pituitary of mice were significantly increased (NTP 1999). Studies on the genotoxicity of ethylbenzene have generally shown a lack of genetic effects (IARC 2000).

Hexadecane

Hexadecane is one of the top 10 constituents of JP-8 (by weight percentage). It partially inhibited the mouse skin carcinogenicity of benzo[*a*]pyrene when it was applied to the skin 3 times/wk with a low dose of benzo[*a*]pyrene (Van Duuren and Goldschmidt 1976).

Naphthalene

Naphthalene is present in JP-8 at about 1.14 wt %. It has been used as a biomarker of JP-8 exposure in that it is detectable in breath, blood, and urine of people exposed to JP-8. The International Agency for Research on Cancer (IARC) has classified napthalene as a possible human carcinogen (Group 2B) on the basis of sufficient evidence of carcinogenicity in experimental animals and inadequate evidence in humans (IARC 2002). Two-year inhalation carcinogenicity studies in B6C3F$_1$ mice and F344/N rats of both sexes were conducted by the NTP. The NTP found clear evidence of the carcinogenicity of naphthalene in male and female F344/N rats exposed to naphthalene vapors at 0, 10, 30, or 60 ppm for 6 h/day, 5 days/wk for 2 yr on the basis of increased incidences of respiratory epithelial adenoma (males, control, low-dose, middle-dose, and high-dose groups, 0%, 12%, 17%, and 31%, respectively; females, 0%, 0%, 8%, and 4%, respectively) and olfactory epithelial neuroblastoma of the nose (males, 0%, 0%, 8%, and 6%; females, 0%, 4%, 6%, and 24%) (Abdo et al. 2001; NTP 2000). The lowest exposure concentration used in the rat studies equals the threshold limit value for the 8-hr, time-weighted average established by the American Conference of Governmental Industrial Hygienists (ACGIH 1999). One male each in the 30- and 60-ppm groups had metastases of olfactory epithelial neuroblastoma to the lungs. Olfactory epithelial neuroblastoma and respiratory epithelial adenoma are unusual in the F344/N rat and had not been observed previously in NTP studies. Neuroblastomas of the nasal olfactory epithelium are rare neoplasms in rodents and humans (Pino et al. 1999; McElroy et al. 1998, as cited by Abdo et al. 2001).

In the B6C3F$_1$ mouse studies, animals were exposed to naphthalene vapors at 0, 10, or 30 ppm 6 hr/day, 5 days/wk for 2 yr, and an increased incidence of alveolar/bronchiolar adenoma was observed in the 30-ppm group of female mice (NTP, 1992, Abdo et al, 1992).

As reviewed by the NTP (2000), naphthalene has been shown to cause sister chromatid exchanges (SCEs) and chromosomal aberrations in Chinese hamster ovary cells, micronuclei (MNs) in human lymphoblastoid MCL-5 cells, and somatic mutations and recombination in *Drosophila*. Naphthalene was not mutagenic in *Salmonella*, nor did it induce DNA damage in *E. coli* (as reviewed by NTP 2000).

Tetradecane

Tetradecane is one of the top 10 constituents of JP-8 (by weight percentage). It exhibited cocarcinogenicity by enhancing the mouse skin carcinogenicity of benzo[*a*]pyrene (when applied to the skin 3 times/wk with a low dose of benzo[*a*]pyrene) (Van Duuren and Goldschmidt 1976) and of ultraviolet radiation (Bingham and Nord 1977). Tetradecane was also tested as a tumor promoter in a two-stage carcinogenesis assay and found to have tumor-promoting activity (Van Duuren and Goldschmidt 1976).

1,2,4,5-Tetramethylbenzene

1,2,4,5-Tetramethylbenzene is one of the top 10 constituents of JP-8 (by weight percentage). No data on its carcinogenicity were identified in the published literature. It was not mutagenic in *Salmonella*, and it did not induce MNs in the in vivo mouse bone marrow cell assay (Janik-Spiechowicz and Wyszynska 1999). It did induce SCEs in the bone marrow of mice in a dose-dependent manner (Janik-Spiechowicz and Wyszynska 1999).

Undecane

Undecane is one of the top 10 constituents of JP-8 (by weight percentage). It exhibited cocarcinogenicity by enhancing the mouse skin carcinogenicity of benzo[*a*]pyrene (when applied to the skin 3 times/wk together with a low dose of benzo[*a*]pyrene) (Van Duuren and Goldschmidt 1976). Undecane was not mutagenic in *Salmonella* in the presence or absence of metabolic activation (Connor et al. 1985).

CONCLUSIONS AND RECOMMENDATIONS

The carcinogenicity of JP-8 has not been investigated in epidemiologic studies. Chronic lifetime inhalation-exposure studies have not been conducted in experimental animals to determine the carcinogenicity of JP-8 or related jet fuels. No increase in the incidence of tumors was observed in 90-day continuous inhalation-exposure studies of JP-5 conducted in F344 rats and C57BL/6 mice (with a 19- or 21-mo observation period after cessation of exposure). Positive results of in vitro genotoxicity tests in cultured human and rat cell lines suggest that JP-8 has the potential to induce DNA damage; however, the genotoxicity of JP-8 has not been evaluated adequately in vivo. As described in Chapter 3, JP-8 is a complex chemical mixture that comprises about 1,000 components. Among those on which carcinogenicity data are available, three chemicals (benzene, ethylbenzene, and naphthalene), which together make up 1% or less (volume/volume) of the fuel, are known to be carcinogenic. The carcinogenicity data available on mixtures similar to JP-8 (such as other jet fuels and MDFs) indicate that most of these materials induce skin tumors in mice when topically applied in excessive amounts. The mixtures have also been shown to have tumor-promoting but not tumor-initiating activity in the two-stage mouse skin tumor model. However, those carcinogenic effects are observed only under conditions of excessive skin irritation.

The subcommittee concludes that the available data are insufficient to draw a conclusion regarding the carcinogenicity of inhaled JP-8. However, because some studies show that chronic dermal exposure to high doses of jet fuels or other petroleum products produces skin tumors, the subcommittee recommends that the Department of Defense (DOD) conduct lifetime carcinogenicity bioassays by the inhalation route in two animal species to determine whether JP-8 is carcinogenic via inhalation. The subcommittee also recommends that DOD follow a cohort of military personnel (including obtaining their exposure history) to determine whether exposure to JP-8 is associated with an increased incidence of various types of cancers.

The subcommittee is aware that Air Force personnel engaged in particular jobs (such as fuel-cell workers) are sometimes dermally exposed to substantial amounts of JP-8 (see Chapter 2). The subcommittee recommends that appropriate protective clothing be worn to avoid dermal exposures to JP-8.

REFERENCES

Abdo, K.M., S. Grumbein, B.J. Chou, and R. Herbert. 2001. Toxicity and carcinogenicity study in F344 rats following 2 years of whole-body exposure to naphthalene vapors. Inhal. Toxicol. 13(10):931-950.

Abdo, K.M., S.L. Eustis, M. McDonald, M.P. Jokinen, B. Adkins Jr, and J.K. Haseman. 1992. Naphthalene: A respiratory tract toxicant and carcinogen for mice. Inhal. Toxicol. 4(4):393-409.

ACGIH (American Conference of Governmental Industrial Hygenists). 1999. 1999 TLVs and BEIs: Threshold Limit Values for Chemical Substances and Physical Agents. Biological Exposure Indices. Cincinnati, OH: ACGIH.

API (American Petroleum Institute). 1988. Lifetime Dermal Carcinogenesis Bioassay of Refinery Streams in C3H/HeJ Mice (API 135r). API Med. Res. Publ. 36-31364. Washington, DC: American Petroleum Institute.

API (American Petroleum Institute). 1989. Short-Term Dermal Tumorigenesis Study of Selected Petroleum Hydrocarbons in Male CD-1 Mice: Initiation and Promotion Phases. Final Report. API 36-32643. Washington, DC: American Petroleum Institute.

ATSDR (Agency for Toxic Substances and Disease Registry). 1998. Toxicological Profile for Jet Fuels (JP-5 and JP-8). U.S. Department of Health and Human Services, Public Health Service, Agency for Toxic Substances and Disease Registry, Atlanta, GA.

Biles, R.W., R.H. McKee, S.C. Lewis, R.A. Scala, and L.R. DePass. 1988. Dermal carcinogenic activity of petroleum-derived middle distillate fuels. Toxicology 53(2-3):301-314.

Bingham, E., and H.L. Falk. 1969. Environmental carcinogens. The modifying effect of cocarcinogens on the threshold response. Arch. Environ. Health 19(6):779-783.

Bingham, E., and P.J. Nord. 1977. Cocarcinogenic effects of n-alkanes and ultraviolet light on mice. J. Natl. Cancer Inst. 58(4):1099-1101.

Bruner, R.H., E.R. Kinkead, T.P. O'Neill, C.D. Flemming, D.R. Mattie, C.A. Russell, and H.G. Wall. 1993. The toxicologic and oncogenic potential of JP-4 jet fuel vapors in rats and mice: 12-month intermittent inhalation exposures. Fundam. Appl. Toxicol. 20(1):97-110.

Clark, C.R., M.K. Walter, P.W. Ferguson, and M. Katchen. 1988. Comparative dermal carcinogenesis of shale and petroleum-derived distillates. Toxicol. Ind. Health 4(1):11-22.

Connor, T.H., J.C. Theiss, H.A. Hanna, D.K. Monteith, and T.S. Matney. 1985. Genotoxicity of organic chemicals frequently found in the air of mobile homes. Toxicol. Lett. 25(1):33-40.

Easley, J.R., J.M. Holland, L.C. Gipson, and M.J. Whitaker. 1982. Renal toxicity of middle distillates of shale oil and petroleum in mice. Toxicol. Appl. Pharmacol. 65(1):84-91.

Freeman, J.J., T.M. Federici, and R.H. McKee. 1993. Evaluation of the contribution of chronic skin irritation and selected compositional parameters to the tumorigenicity of petroleum middle distillates in mouse skin. Toxicology 81(2):103-112.

Gaworski, C.L., J.D. MacEwen, E.H. Vernot, R.H. Bruner, and M.J. Cowan Jr. 1984. Comparison of the subchronic inhalation toxicity of petroleum and oil shale JP-5 jet fuels. Pp. 33-48 in Advances in Modern Environmental Toxicology, Vol. 6. Applied Toxicology of Petroleum Hydrocarbons, H.N. MacFarland, C.E. Holdworth, J.A. MacGregor, R.W. Call, and M.L. Lane, eds. Princeton, NJ: Princeton Scientific Publishers.

Gaworski, C.L., J.D. MacEwan, E.H. Vernot, C.C. Haun, H.F. Leahy, R.H. Bruner, G.B. Baskin and M.J. Cowan Jr. 1985. Evaluation of the 90-day Inhalation Toxicity of Petroleum and Oil Shale JP-5 Jet Fuels. AFAMRL-TR-85-035. NMRI 85-18. Air Force Aerospace Medical Research Laboratory, Wright Patterson Air Force Base, OH. April 1985.

HSDB (Hazardous Substance Data Bank). 2001. Tertiary-butylbenzene. Hazardous Substance Data Bank, National Library of Medicine. [Online]. Available: http://toxnet.nlm.nih.gov/ [May 31, 2002].

IARC (International Agency for Research on Cancer). 1987. Pp. 120-222 in Overall Evaluation of Carcinogenicity: An Updating of IARC Monographs Volumes 1 to 42. IARC Monographs on the Evaluation of the Carcinogenic Risk to Humans, Suppl. 7. Lyon: IARC.

IARC (International Agency for Research on Cancer). 2000. Ethylbenzene. Pp. 227-266 in Some Industrial Chemicals, IARC Monographs on the Evaluation of the Carcinogenic Risk to Humans, Vol. 77. Lyon: IARC.

IARC (International Agency for Research on Cancer). 2002. Some Traditional Herbal

Medicines, Some Mycotoxins, Naphthalene and Styrene, IARC Monographs on the Evaluation of the Carcinogenic Risk to Humans, Vol. 82 (in preparation, as cited at http://monographs.iarc.fr/htdocs/announcements/vol82.ht).

Janik-Spiechowicz, E., and K. Wyszynska. 1999. Genotoxicity evaluation of tetramethylbenzenes. Mutat. Res. 439(1):69-75.

Lewis, S.C., R.W. King, S.T. Cragg, and D.W. Hillman. 1984. Skin carcinogenic potential of petroleum hydrocarbons: Crude oil, distillate fractions and chemical class subfractions. Pp 139-150 in Advances in Modern Environmental Toxicology, Vol. 6. Applied Toxicology of Petroleum Hydrocarbons, H.N. MacFarland, C.E. Holdworth, J.A. MacGregor, R.W. Call, and M.L. Lane, eds. Princeton, NJ: Princeton Scientific Publishers.

Mattie, D.R., C.L. Alden, T.K. Newell, C.L. Gaworski, and C.D. Flemming. 1991. A 90-day continuous vapor inhalation toxicity study of JP-8 jet fuel followed by 20 or 21 months of recovery in Fischer 344 rats and C57BL/6 mice. Toxicol. Pathol. 19(2):77-87.

Mattie, D.R., G.B. Marit, C.D. Flemming, and J.R. Cooper. 1995. The effects of JP-8 jet fuel on male Sprague-Dawley rats after a 90-day exposure by oral gavage. Toxicol. Ind. Health 11(4):423-435.

McElroy, E.A., Jr., J.C. Buckner, and J.E. Lewis. 1998. Chemotherapy for advanced esthesioneuroblastoma: The Mayo Clinic experience. Neurosurgery 42(5):1023-1028.

McKee, R.H., R.T. Plutnick, and R.T. Przygoda. 1989. The carcinogenic initiating and promoting properties of a lightly refined paraffinic oil. Fundam. Appl. Toxicol. 12(4): 748-756.

Nessel, C.S. 1999. A comprehensive evaluation of the carcinogenic potential of middle distillate fuels. Drug Chem. Toxicol. 22(1):165-180.

Nessel, C.S., J.J. Freeman, R.C. Forgash, and R.H. McKee. 1999. The role of dermal irritation in the skin tumor promoting activity of petroleum middle distillates. Toxicol. Sci. 49(1):48-55.

Nessel, C.S., R.A. Priston, R.H. McKee, G. Cruzan, A.J. Riley, R. Hagemann, R.T. Plutnick, and B.J. Simpson. 1998. A comprehensive evaluation of the mechanism of skin tumorigenesis by straight-run and cracked petroleum middle distillates. Toxicol. Sci. 44(1):22-31.

Nevada State Health Division. 2002. Churchill County (Fallon) Childhood Leukemia Update, August 20, 2002 Community Meeting. [Online] Available : http://health2k.state.nv.us/healthofficer/Leukemia/Fallon.htm [October 28, 2002].

NRC (National Research Council). 1996. Permissible Exposure Levels for Selected Military Fuel Vapors. Washington, DC: National Academy Press.

NTP (National Toxicology Program). 1986. Toxicology and Carcinogenesis Studies of Marine Diesel Fuel and JP-5 Navy Fuel (CAS No. 8008-20-6) in B6C3F1 Mice (Dermal Studies). NTP 310. NIH 86-2566. Research Triangle Park, NC: National Toxicology Program/National Institutes of Health.

NTP (National Toxicology Program). 1992. NTP Technical Report on the Toxicology and Carcinogenesis Studies of Naphthalene (CAS No. 91-20-3) in $B6C2F_1$ Mice (Inhalation Studies). NTP TR 410. NIH 92-3141. Research Triangle Park,

NC: U.S. Dept. of Health and Human Services, Public Health Service, National Institutes of Health.

NTP (National Toxicology Program). 1999. TR-466. Toxicology and Carcinogenesis Studies of Ethylbenzene (CAS No. 100-41-4) in F344/N Rats and $B6C_3F_1$ Mice (Inhalation Studies) NTIS: PB99-134694. [Online] Available: http://ntp-server.niehs.nih.gov/htdocs/LT-studies/tr466.html [accessed November 21, 2002].

NTP (National Toxicology Program). 2000. NTP Technical Report on the Toxicology and Carcinogenesis Studies of Naphthalene (CAS No. 91-20-3) in F344/N Rats (Inhalation Studies). NTP TR 500. NIH 01-4434. Research Triangle Park, NC: U.S. Department of Health and Human Services, Public Health Service, National Institutes of Health.

Parent, M.E., Y. Hua, and J. Siemiatycki. 2000. Occupational risk factors for renal cell carcinoma in Montreal. Am. J. Ind. Med. 38(6):609-618.

Pino, M.V., M.G. Valerio, G.K. Miller, J.L. Larson, D.L. Rosolia, Z. Jayyosi, C.N. Crouch, J.Q. Trojanowski, and L.E. Geiger. 1999. Toxicologic and carcinogenic effects of the type IV phosphodiesterase inhibitor RP 73401 on the nasal olfactory tissue in rats. Toxicol. Pathol. 27(4):383-394.

Selden, A., and G. Ahlborg Jr. 1991. Mortality and cancer morbidity after exposure to military aircraft fuel. Aviat. Space Environ. Med. 62(8):789-794.

Siemiatycki, J., R. Dewar, L. Nadon, M. Gerin, L. Richardson, and S. Wacholder. 1987. Associations between several sites of cancer and twelve petroleum-derived liquids. Scand. J. Work Environ. Health 13(6):493-504.

Skisak, C. 1991. The role of chronic acanthosis and subacute inflammation in tumor promotion in CD-1 mice by petroleum middle distillates. Toxicol. Appl. Pharmacol. 109(3):399-411.

Van Duuren, B.L., and B.M. Goldschmidt. 1976. Cocarcinogenic and tumor-promoting agents in tobacco carcinogenesis. J. Natl. Cancer Inst. 56(6):1237-1242.

Appendix A

Permissible Exposure Levels for Selected Military Fuel Vapors: Contents and Executive Summary (NRC 1996)

Permissible Exposure Levels for Selected Military Fuel Vapors

SUBCOMMITTEE ON PERMISSIBLE EXPOSURE LEVELS FOR MILITARY FUELS

COMMITTEE ON TOXICOLOGY

BOARD ON ENVIRONMENTAL STUDIES AND TOXICOLOGY

COMMISSION ON LIFE SCIENCES

NATIONAL RESEARCH COUNCIL

NATIONAL ACADEMY PRESS
WASHINGTON, D.C., 1996

Contents

Executive Summary

The U.S. Navy is in the final stages of designing a strategic sealift ship to transport already-fueled vehicles—armored tanks, tanker trucks, other trucks of various sizes, trailers, jeeps, and helicopters. Prefueling will eliminate the need for fueling at docking and will permit deployment of the vehicles as soon as they are unloaded from the ship. All the military vehicles transported on the ship are designed to use jet-propulsion (JP) fuels JP-5 or JP-8 to avoid the need for different fuels. Diesel fuel marine (DFM) is used to operate the ship.

The Navy's Occupational Safety and Health Standards Board has considered the potential for storage and operation of the fueled vehicles in the ship's cargo holds to be hazardous to naval service personnel exposed to fuel vapors during the servicing of these vehicles or while working in their vicinity. To protect personnel from exposures to toxic concentrations of fuel vapors, the board recommended an interim 8-hr time-weighted average (TWA) permissible exposure limit (PEL) of 350 mg/m^3 and a 15-min short-term exposure limit (STEL) of 1,800 mg/m^3 for vapors from all three fuels. Those interim exposure limits were based on the board's review of the manufacturers' technical documentation and the National Institute of Occupational Safety and Health's recommendations for maximum exposure to refined petroleum solvents.

The Navy decided that an independent review of the proposed interim exposure limits would be useful and requested that the National Research Council (NRC) perform the following tasks: (1) review the toxicity data on these fuels, (2) determine the adequacy of the Navy's proposed PELs and STELs, and (3) recommend changes, if needed, to the proposed limits.

In response to the Navy's request, the NRC assigned this project to the Committee on Toxicology (COT). The COT convened the Subcommittee on Permissible Exposure Levels for Military Fuels,[1] which prepared this report. The subcommittee based its evaluation of the Navy's interim PELs and STELs on a detailed examination of current data on the toxicity of fuel vapors from JP-5, JP-8, JP-4, and DFM in animals and humans. JP-4 is included in this analysis because more information is available on the toxicity of JP-4 vapors than on the other three fuels and because the composition of JP-4 vapors is sufficiently similar to those of JP-5 and JP-8 vapors. The toxicity of the vapors from all these fuels is expected to be similar.

The subcommittee did not address the potential toxicity resulting from exposure to respirable aerosols of the total fuels. If the Navy finds evidence of potential for exposures to respirable aerosols, which are much more toxic, new recommendations for limiting such exposures will need to be developed. It is understood that the board intends to use the exposure levels for the three fuels to help determine ventilation requirements in the cargo holds of the new strategic sealift ships and to prevent toxic exposures of service personnel to these fuels.

ADVERSE EFFECTS

The major adverse effects resulting from inhalation of these military fuel vapors are kidney, central-nervous-system (CNS), liver, and carcinogenic effects.

Kidney

The renal toxicity of these military fuels was studied in rats and mice of both sexes. Adverse effects in the kidneys were observed only in male rats after inhalation exposure. Histological sections from the kidneys of affected animals were examined, and the presence of the characteristic hyaline droplets, suggestive of an $\alpha 2u$-globulin pathogenesis, was confirmed. Current scientific thinking is that these findings are not relevant to humans because this kidney lesion appears to be unique to the male rat.

---

[1]The term "military fuels" in this report refers only to JP-5, JP-8, JP-4, and DFM vapors.

Central Nervous System

In one epidemiological investigation, 30 workers exposed to jet fuel at a Swedish jet-motor factory for an average of 17 years were studied for possible adverse health effects. The TWA exposure concentrations from one-time measurements of workers in different job categories were calculated to be 420 mg/m^3 for component testers, 130 mg/m^3 for engine testers, and 190-250 mg/m^3 for mechanics. The overall TWA concentration from one-time measurements was 300 mg/m^3; peak exposures ranged from approximately 1,200 to 3,200 mg/m^3. Significant differences between the exposed and nonexposed workers were found with respect to CNS effects. The majority of the exposed workers reported acute symptoms of dizziness, headache, nausea, and fatigue. Chronic symptoms included greater incidence of neurasthenic symptoms (depressed mood, lack of initiative, sleep disturbances, memory impairment, headache, dizziness, and fatigue). The exposed workers also showed higher performance degradation in a variety of performance tests than the nonexposed workers. The neurophysiological examination with electroencephalograms showed greater incidence of abnormalities in jet-fuel exposed workers than in nonexposed workers. However, the findings of CNS effects attributable to long-term exposure were considered questionable for a number of reasons, including weak and inconsistent evidence of impairment, inadequate methods of evaluation, inadequate consideration of confounding factors, a small cohort of workers,and a lack of quantitative information on exposure.

Liver

Several investigators have studied the effects of subchronic exposure to military fuel vapors on the liver in experimental animals. No liver histopathological changes were found in three inhalation studies in which animals were exposed intermittently (occupational-type exposure) to fuel vapors. In one study, rats exposed to JP-5 vapors at concentrations of 1,000 or 1,600 mg/m^3 for 6 hr per day, 5 days per week for 6 weeks showed no evidence of adverse effects on the liver. In the second study, rats, mice, dogs, and monkeys exposed to JP-4 vapors at 2,500 or 5,000 mg/m^3 for 6 hr per day, 5 days per week for 8 months showed no evidence of exposure-related effects except a slight increase in liver weight in the female rats. In the third study, rats and mice exposed to JP-4 vapors at 1,000 or 5,000 mg/m^3 for 6 hr per day, 5 days per week for 12 months showed no liver toxicity. No clear evidence of hepatic neoplasia in rats or mice was found. Based on this study, the subcommittee identified a no-observed-adverse-effect level (NOAEL) of 5,000 mg/m^3, which was used to calculate the PEL. By dividing the NOAEL of 5,000 mg/m^3 by an uncertainty factor of 10 for interspecies extrapolation, the PEL was calculated to be 500 mg/m^3. No uncertainty factor for intraspecies variation was applied because the exposed Navy personnel are considered to be healthy.

Carcinogenicity

Carcinogenicity of military fuels has been studied in humans and animals. An epidemiological study of approximately 2,200 Swedish military personnel exposed to jet-fuel vapors at concentrations greater than 350 mg/m^3 for several years did not show increased incidence of cancer. It should be noted that this study was only capable of detecting high risks of cancer because there were few cancer deaths, the sample was small, and the follow up was short.

Several studies of petroleum workers, ranging from refinery workers to service-station attendants, reported increases in cancer, but few studies reported on persons exposed only to jet-fuel vapors. Exposure to benzene appears to be of consequence in many of the excesses found. In long-term animal studies involving inhalation exposure to unleaded gasoline, kidney cancers were observed only in male rats. That finding raises the question of whether longer exposure of male rats to JP-5, JP-8, or DFM might also result in increased kidney cancers. However, the increased incidence of kidney cancer in male rats exposed to the gasoline was due to an α2u-globulin nephropathy—a lesion that apparently does not occur in humans, in other animals, or in female rats.

Based on the available human and animal data, the subcommittee concluded that inhalation of JP-5, JP-8, and DFM vapors does not present a carcinogenic risk to humans. That conclusion is supported by studies that show that these military fuels are not genotoxic. However, laboratory studies provided evidence of potential carcinogenicity of DFM via the dermal route. Epidemiological studies show skin-cancer excesses in certain industrial workers, such as machine operators, whose skin might come into contact with lubricating oils derived from coal tar or petroleum. Exposure conditions in the studies that resulted in excessive skin damage are unlikely to occur on Navy ships.

CONCLUSIONS AND RECOMMENDATIONS

The American Conference of Governmental Industrial Hygienists has not recommended exposure limits for the military fuels that are the subject of this report. The Occupational Safety and Health Administration and other regulatory agencies also have not promulgated standards for these fuels.

The toxicity data on military fuels are sparse. No reliable information was found to indicate a need to change the Navy's proposed PEL of 350 mg/m^3. The findings in Swedish jet-motor factory workers of chronic CNS effects—performance degradation and neurasthenic symptoms—attributable to long-term exposure to jet fuels at TWA concentrations of 300 mg/m^3 were considered questionable for reasons discussed above. The studies of hepatotoxicity in experimental animals were also considered to be of questionable significance. The PEL of 500 mg/m^3 was based on a slight increase in liver weight in rats—an effect that was reversible and not accompanied by any histopathological change. Based on the available information from studies in humans and animals and based on expert judgment, the subcommittee concludes that the Navy board's 8-hr PEL of 350 mg/m^3 for JP-5, JP-8, and DFM is adequate to protect the health of naval personnel occupationally exposed to military fuels. Due to the uncertainty surrounding (1) the CNS effects observed in Swedish jet-motor factory workers from chronic exposure to jet fuel at TWA concentrations of 300 mg/m^3 and (2) the NOAEL of 500 mg/m^3 derived from liver toxicity studies in rats and mice, the subcommittee recommends that the PEL of 350 mg/m^3 be considered interim until further research is completed.

Data needed to evaluate the adequacy of the Navy's 15-min STEL of 1,800 mg/m^3 for the three fuels are sparse. The subcommittee considered the acute CNS effects (e.g., dizziness, headache, nausea, and fatigue) in the Swedish jet-motor factory workers to be the most critical health effects for determining the adequacy of the STELs. Based on the limited information on exposure concentrations and the attribution of CNS symptoms to peak exposures of approximately 1,000 mg/m^3 or higher, the subcommittee recommends that the Navy's current STEL be lowered from 1,800 mg/m^3 to 1,000 mg/m^3 to avoid acute CNS toxicity. The STEL of 1,000 mg/m^3 should also be considered an interim recommendation until further research is completed.

The subcommittee also recommends the following:

- Appropriate protective clothing should be worn to reduce dermal exposure because of the evidence of the carcinogenic potential of DFM via the dermal route.
- Because respirable aerosols of military fuels are much more toxic than vapors, naval personnel should avoid exposure to aerosolized fuel. If a potential for exposure to aerosolized fuel exists, protective clothing and respiratory equipment should be worn.
- The Navy should complete the following research to improve its ability to assess the health risks associated with the use of military fuels:

—Obtain information on exposures occurring during operational procedures, including exposures to respirable aerosols of unburned fuels. Samples should be talon in the breathing zone of the service personnel. Breath analysis of exposed personnel is recommended to determine the extent of individual exposures to fuel vapors.

—Conduct studies on the possible effects of high-level acute and low-level chronic exposure to military fuel vapors on the CNS, including the effects on the performance of personnel. At present, very little information exists. Anecdotal accounts do not provide adequate documentation of exposures leading to reduced performance.

—Conduct further research on the effect of military fuel vapors on hepatotoxicity in experimental animals; this research would help to identify the NOAEL with greater confidence.

Appendix B

JP-8 Final Risk Assessment: Contents, Executive Summary, and Introduction (TIEHH 2001)[1]

[1]Reprinted with permission from the U.S. Air Force.

The Institute of Environmental and Human Health
(TIEHH)
1207 Gilbert Drive, Lubbock, Texas 79409

JP8
Final Risk Assessment

August 2001

Table of Contents

EXECUTIVE SUMMARY

The following report represents the final report of preliminary results of the protocol to assess the health and performance effects of acute exposure to Jet Fuel number 8. Texas TechUniversity, Institute of Environmental and Human Health, in conjunction with the United StatesAir Force, hosted this protocol with funding from Strategic Environmental Research and Development Program. Additional collaborators include the University of Cincinnati, and the Oregon Health Sciences University, the University of Texas, the University of North Carolina,Johns Hopkins University, the US Navy Toxicology Laboratory at Wright-Patterson AFB, OH,the NIOSH, and EPA/NERL herein referred to as the JP-8 Research Team.

Jet Propellant type 8 (JP8) jet fuel is the recognized battlefield fuel for all military operations for the United States, well beyond the year 2025, and represents the single largest source of chemical exposure to Department of Defense (DOD) personnel. Currently, DOD and its NATO partners use approximately 5 billion gallons of JP8 annually. The commercial equivalent, Jet-A, is the primary jet fuel used by aircraft in the US. Worldwide use of kerosene-based jet fuel is over 58 billion gallons per year.

The study was conducted at multiple Air Force installations. Dyess AFB, TX, served as the beta test site for participant selection, specimen collection, and exposure assessment. The lessons learned from the Dyess AFB beta test allowed the JP8 Research Team to improve data collection processes and study logistics, thus reducing the operational study impact at other Air Force bases involved in the study. Data was collected at the following sites: Davis Monthan AFB, AZ, Seymour Johnson AFB, NC, Langley AFB, VA, Pope AFB, NC, Little Rock AFB, AR, and Hurlbert Field, FL. Specimens and data collected from these locations were analyzed at established laboratory facilities operated by the universities and government agencies involved in the study.

Two groups of airmen were enrolled. Those designated as **JP8 exposed** consisted of active duty Air Force personnel who routinely worked with or are exposed to JP8 in the performance of their duties. Most exposed volunteers worked in Aircraft Fuel Cell Maintenance shops. These workers routinely performed maintenance activities requiring entry into aircraft fuel tanks. Other exposed volunteers worked in either the Fuels Specialty or Fuels Transportation shops. In order to qualify for the study, exposed volunteers were required to have least 9 months of persistent exposure to jet fuel (such as fuel tank entry at least one hour twice weekly).

Unexposed volunteers were intended to represent the population of active duty Air Force personnel assigned to the Air Force installation where the study was being conducted. They consisted of active duty personnel assigned to the same Air Force installation as the JP8 exposed volunteers, but who do not have routine contact with JP8 or other fuels during the performance of their duty. A wide variety of job classifications were represented in the unexposed group. Since nearly all JP8 exposed volunteers were enlisted personnel, attention was paid to ensure, for the most part, that Air Force officers were not selected to participate. In rare cases, officers were included where the researchers felt their inclusion would not bias the analysis.

Broadly, JP8 exposure was measured both externally in the environment immediately surrounding enrolled workers and internally through the use of several body burden measures. The impact of exposure was evaluated using a series of neurological, hormonal and immunological measures. Cytotoxic and genotoxic effects of JP8 exposure were also evaluated. The activity of Glutathione-S-Transferase (G-S-T), a gene-regulated enzyme associated with increased susceptibility to multiple oxidative stressors including jet fuel and linked to adverse health outcomes, was also measured. Self-reported health problems, health care visit frequency, and early indicators of liver and kidney damage were investigated as part of the study.

The characterization of JP8 health risks, conducted by Texas Tech University, and the identification of uncertainties accounted for exposure measures of JP8 and measures of effect. The preliminary risk characterization attempted to determine the association between the various measures of effect used in this study and assesses the overall impact, by JP8 dose, on workers exposed to the fuel.

INTRODUCTION

Risk Assessment of Acute Exposure to Jet Fuel

Jet Propellant type 8 (JP8) jet fuel is the recognized battlefield fuel for all military operations for the United States, well beyond the year 2025, and represents the single largest source of chemical exposure to Department of Defense (DOD) personnel. Currently, DOD and its NATO partners use approximately 5 billion gallons of JP8 annually. The commercial equivalent, Jet-A, is the primary jet fuel used by aircraft in the US. Worldwide use of kerosene-based jet fuel is over 58 billion gallons per year.

Over the past 20 years, JP8 largely replaced JP4 as the primary aircraft fuel for US military aircraft. JP4, which is chemically similar to gasoline, is highly volatile. Explosive fires in both occupational and operational settings were experienced in military aircraft powered by JP4. JP8, although chemically similar to kerosene, is much less volatile. It is a much safer fuel to handle and less likely to propagate an explosion during instances when military aircraft fuel tanks suffer artillery or small arms damage during operational situations.

As JP8 was phased into the military inventory, exposed personnel began voicing concerns about the potential health effects of exposure. Aircraft groundcrew members reported objectionable odors, skin irritation, dizziness and the persistent taste of jet fuel long after exposure. These concerns prompted the Air Force Surgeon General to task the Air Force Institute for Environment, Safety and Occupational Health Risk Analysis (AFIERA) and the Air Force Research Laboratory (AFRL) to address personal exposure and toxicological hazards from JP8.

A reference report by the Center for Disease Control and Prevention's (CDC) Agency for Toxic Substances and Disease Registry (ATSDR), "Toxicology Profile for Jet Fuels (JP-5 and JP8)," in 1997 indicated that the toxicities of jet fuel and their mechanisms are not well-defined. According to ATSDR, data gaps exist on dose-response, reproductive system, developmental effects, immune system, neurological system, biomarkers of exposure and effect, rates of absorption, distribution and excretion of, and toxicokinetics in current research of human health effects from jet fuel exposure. Recently, JP8 jet fuel was selected as a priority hazardous chemical requiring establishment of an acute exposure limit by the Environmental Protection Agency's (EPA) National Advisory Committee for Acute Exposure Guidelines for Hazardous Substances (NAC-AEGL), a subcommittee of the Office of Pollution Prevention and Toxics, US EPA. The NAC-AEGL further identified data gaps in the toxicology profile of jet fuel as submitted by the ATSDR. Recommendations from the NAC-AEGL include measuring total body burden, identifying biomarkers of exposure, conducting an epidemiology study of worst-case exposed populations, conducting neurological assessment, establishing reference dose (RfD) and risk assessment of exposure from JP8. In addition, a 1996 report by the National Research Council's Committee on Toxicology (COT) identified data gaps in occupational exposure assessments, breath analysis, quantitative neurological effects and hepatotoxicity.

The COT report recommended the following:

a. Obtain information on exposures during operational procedures, including exposures to respirable aerosols of unburned fuels.

b. Conduct studies on the possible effects of high-level acute and low-level chronic exposures to military fuel vapors on CNS, including the effects on performance of military personnel.

c. Conduct further research on the effect of military fuel vapors on hepatotoxicity in experimental animals.

Based on the Air Force Surgeon General's tasking and ATSDR, COT and NAC-AEGL recommendations, AFIERA initiated a program to evaluate all environmental, safety and occupational health aspects of jet fuel and began collaboration with the National Institute for Occupational Safety and Health (NIOSH), the Environmental Protection Agency, National Exposure Research Laboratory (EPA-NERL), the National Institute for Environmental Health Sciences (NIEHS), and selected academic institutions to resolve open issues regarding JP8. The USAF JP8 Environmental, Safety and Occupational Health Integrated Process Team (IPT), formed in 1996 and in coordination with the Air Force Office of Scientific Research (AFOSR), conducted and funded animal toxicology studies on aerosol exposure, dermal flux and adsorption, biomarkers and neurological assessments. The IPT has also conducted and funded occupational exposure studies to include ambient vapor and aerosol exposure assessment, breath sampling, and heat stress assessment.

Based on exposure data from previous AFIERA studies, fuel tank repair operations of single-point-entry fuel bladders containing fire suppressant foam were determined as the worst-case exposure situations. The highest exposure results were measured in operations performed inthe C-130 Hercules transport aircraft's auxiliary fuel tanks.

The studies conducted by AFIERA and other investigators, including toxicology studies supported by the AFOSR, validated the need for research on JP8 impact on workers in occupational settings. In particular, studies of the acute effects of exposure were considered most important. Based on these assessments, this study, entitled **Risk Assessment of Acute Exposure to Jet Fuel**, was developed and initiated. Prior to this study, no occupational exposure cohort studies had been conducted to assess the effects from acute exposure to JP8 jet fuel. Further, no acute exposure or risk assessment studies had attempted to link quantitative neurological measurements to ambient exposure, biomarkers, and total body burden. This study breaks new ground by correlating ambient exposure with human body burden and neurological performance measures. The results of this study are intended to aid in establishing limits for exposure in both occupational and community settings. The study helps to determine specific occupational exam requirements, personal protective equipment requirements and methods for monitoring exposure. Additionally, by correlating ambient exposure measures with health and performance outcomes, we hope to use the data obtained from this study to extrapolate the extent of community risks associated with ubiquitous, low-level jet fuel exposure.

The study was conducted in conjunction with Texas Tech University, the University of Cincinnati, and the Oregon Health Sciences University. Additional collaborators include the University of Texas, the University of North Carolina, Johns Hopkins University, the US Navy Toxicology Laboratory at Wright-Patterson AFB, OH, the NIOSH, and EPA/NERL.

The study's purpose was to assess the influence of acute exposure to jet fuel on the health, safety and operational capability of the Air Force population and gain insight into the risk posed by JP8 on the general local population.

The specific aims were to

a. Compare exposure levels of a selected worst-case exposed cohort to the generally unexposed base workforce.

b. Determine level of body burden of jet fuel within each exposure group.

c. Analyze biological specimens from each subject group for jet-fuel-linked specific biomarkers of exposure and effect.

d. Perform an epidemiology analysis of each subject group.

e. Assess the impact of JP8 exposure on performance and health outcomes.

f. Perform a risk analysis for environmental and occupational communities based on collected sample data.

The primary hypotheses addressed through this study are the following:

> Is exposure to JP8 detrimental to the health and safety of flightline workers? Does a low-level ambient exposure to jet fuel have an adverse impact on the general community at an Air Force installation?

The study was conducted at multiple Air Force installations. Dyess AFB, TX, served as the beta test site for participant selection, specimen collection, and exposure assessment. The lessons learned from the Dyess AFB beta test allowed the JP8 Research Team to improve data collection processes and study logistics, thus reducing the operational study impact at other Air Force bases involved in the study. Data was collected at the following sites: Davis Monthan AFB, AZ, Seymour Johnson AFB, NC, Langley AFB, VA, Pope AFB, NC, Little Rock AFB, AR, and Hurlbert Field, FL. Specimens and data collected from these locations were analyzed at established laboratory facilities operated by the universities and government agencies involved in the study.

General Methods

The Risk Assessment of Acute Exposure to Jet Fuel study measured JP8 exposures in an operational environment and assessed the impact of exposure on the performance and health of those enrolled in the study. JP8 exposure was measured both externally in the environment immediately surrounding enrolled workers and internally through the use of several body burden measures. The impact of exposure was evaluated using a series of neurological, hormonal and immunological measures. Cytotoxic and genotoxic effects of JP8 exposure were also evaluated. The activity of Glutathione-S-Transferase (G-S-T), a gene-regulated enzyme associated with increased susceptibility to multiple oxidative stressors including jet fuel and linked to adverse health outcomes, was also measured. Self-reported health problems, health care visit frequency, and early indicators of liver and kidney damage were investigated as part of the study.

Study Logistics

The Jet Fuel Research Team, a group of approximately 30 researchers from six academic institutions, two government agencies and two military services, traveled to six Air Force bases in the continental United States to conduct the study. Visits were coordinated in advance to obtain Commander permission to conduct the study. Commanders were briefed in person or by video teleconference prior to the visit to provide information on the rationale for the study, the study goals, milestones to be accomplished during the visit, and the logistics associated with conducting the study on their base. The study was conducted during a two-week period at each study site. One Air Force base was visited every month between April and September 2000. A beta test was conducted prior to the initial site visit to test the logistics of moving people and equipment and synchronizing the timing of multiple specimen collections and testing applications.

Study Subject Recruitment:

Recruitment at each study site was initiated prior to study team arrival and continued throughout the first week of the visit. Subjects were recruited for the study through several vehicles. Since the primary exposure group for the study were workers from shops where contact with jet fuel routinely occurs, the supervisors of such shops as Aircraft Fuel Systems Maintenance, Fuels Transportation, and Fuels Specialty were directly contacted to gain support for the study and solicit volunteers. Members of the fuels community, particularly aircraft fuel systems maintenance personnel, supervisors and commanders, showed high interest in the project and large numbers of workers from these shops volunteered for the study.

Additional recruitment efforts consisted of briefings at Commanders Calls, and informational press releases and solicitation advertisements in local military installation newspapers. At some study locations, First Sergeants were contacted to help gain support for the study. A financial incentive of $50.00 was provided by Texas Tech University to compensate subjects for their participation outside of regular duty hours. Those who completed all requested tests and provided all requested specimens received $50.00. Any subject who dropped out prior to completing the study received $10.00.

Recruitment was successful at all study locations. At several locations, volunteers were turned away after a sufficient number of subjects was achieved. While the study actively recruited females, few women work in jobs where jet fuel exposure occurs. The unexposed to exposed ratio for women was increased 2 : 1 as originally planned 4 : 1 in an attempt to improve the ability to detect differences in effect.

Study Subject Enrollment:

Two groups of airmen were enrolled. Those designated as **JP8 exposed** consisted of active duty Air Force personnel who routinely worked with or are exposed to JP8 in the performance of their duties. Most exposed volunteers worked in Aircraft Fuel Cell Maintenance shops. These workers routinely performed maintenance activities requiring entry into aircraft fuel tanks. Other exposed volunteers worked in either the Fuels Specialty or Fuels Transportation shops. In order to qualify for the study, exposed volunteers were required to have least 9 months of persistent exposure to jet fuel (such as fuel tank entry at least one hour twice weekly).

Unexposed volunteers were intended to represent the population of active duty Air Force personnel assigned to the Air Force installation where the study was being conducted. They consisted of active duty personnel assigned to the same Air Force installation as the JP8 exposed volunteers, but who do not have routine contact with JP8 or other fuels during the performance of their duty. A wide variety of job classifications were represented in the unexposed group. Since nearly all JP8 exposed volunteers were enlisted personnel, attention was paid to ensure, for the most part, that Air Force officers were not selected to participate. In rare cases, officers were included where the researchers felt their inclusion would not bias the analysis.

All volunteers were informed of the nature of the study and the potential risks associated with participation. By groups of approximately 50, volunteers were given a 30 to 45 minute briefing by an occupational medicine physician. The script used for the briefing had undergone extensive review and testing prior to employment. Groups of researchers and potential volunteers were asked to comment on the briefing during the beta-testing portion of the study. In addition, volunteers at each study site were asked to comment of the acceptability and completeness of the briefing. Without exception, the members of the JP8 study team and study volunteers considered the standardized briefing highly acceptable.

In addition to the briefing, study volunteers were asked to complete a questionnaire designed to obtain information on specific criteria that could disqualify them from participating in the study. Exclusion criteria consisted of conditions that would impact the validity of either study effects or exposure measures. Those criteria were as follows:

1. Alcohol use within 24 hours prior to entering the study period
2. Injury requiring medical attention within the last 6 months
3. History of melanoma
4. History of congenital night blindness
5. History of lung or ovarian cancer
6. History of adult cerebral vascular accident
7. History of diabetes
8. History of scoliosis
9. Major visual impairment
10. Clinical diagnosis of seizures
11. On medical profile
12. Pregnancy
13. Currently taking any medications determined by an occupational medicine physician to be disqualifying. Such medications included:

   a. Hypertension medication
   b. Antacids or medication for heartburn
   c. Diet pills or other stimulants
   d. Tranquilizers or muscle relaxants
   e. Antidepressive medication
   f. Psychotherapeutic medication
   g. Large doses of megavitamins containing high levels of antioxidants

Each volunteer underwent a personal interview with either an occupational or preventive medicine board certified physician where the volunteer's completed questionnaire was reviewed and specific volunteer questions were addressed. After the physician determined the volunteer was eligible to participate in the study, the volunteer and the physician completed an informed consent document. The new enrollee was then given appointments for study testing. Each enrollee was assigned a unique study code consisting of the first three letters of his or her assigned Air Force base, e.g. Pope AFB = POP, and a randomly generated number between 1,000 and 9,999. A reference log consisting of enrollee's social security number, subject code, and exposure group classification was created, maintained, and safeguarded by the occupational medicine physician. All researchers throughout specimen collection, performance testing, and data analysis phases of the study used the study codes for recording information relative to the enrollee. The use of study codes helped maintain subject confidentiality and assisted in blinding researchers to enrollee exposure status. At the end of the study, the reference log was forwarded to Texas Tech University for permanent storage.

Specimen and Data Collection

In most cases, all exposure measurements and performance/health effects testing were conducted during one subject's workday. Enrollees typically reported for testing on an appointed morning. Each subject was asked about their alcohol and tobacco consumption during the 24 hours prior to testing and whether he or she was experiencing cold or allergy symptoms. Those with cold or allergy symptoms and those who had consumed alcohol within 24 hours were rescheduled to another day whenever possible. Tobacco use was recorded.

From those who met morning test entry parameters, specimens of blood, breath, urine and epidermal skin were collected. Samples of the cells from the interior of the cheek were also collected for later testing. The enrollees completed a series of tests designed to measure various neurological parameters. Prior to returning to work, each volunteer was fitted with equipment designed to collect samples of the air within their breathing zones during the work period. Enrollees were also fitted with equipment designed to measure their heart rate and core body temperature throughout the workday.

After undergoing morning testing, the enrollees returned to their usual workplace and performed routine duties for a period of at least 4 hours. During the time the enrollees were at work, members of the research team collected environmental measures. While most of the environmental samples were collected in or near the Aircraft Fuel System Maintenance Shop, a representative number of samples were gathered from other locations to ensure that those enrollees categorized as unexposed were, in fact, unexposed to jet fuel or similar chemicals.

In the afternoon, enrollees returned to the study site where environmental and vital status monitoring equipment was removed. Post-workday specimens of blood, breath, urine and epidermis were collected and a series of tests similar to those conducted in the morning were repeated. Questionnaires were applied to the enrollees to obtain information regarding the level of mental and physical exertion experienced during the day and details of the individual's activities throughout the work period. Questionnaires designed to capture information on self-reported symptoms, lifestyle risk (such as smoking and drinking), and the use of personal protective equipment were also applied.

After completing all specimen collection and testing, the enrollees received their study stipend and were released. At the end of the week, the researchers departed the base. Of note, the researchers collected information on the exact time of day each specimen was collected and each test was performed for each enrollee using a subject-time-series log. These time-series data were made available to all investigators to aid in analysis.

One test, the electroretinogram (ERG), was not accomplished during the typical data collection week. The ERG, a method of measuring retinal function, was administered to a subset of enrollees during the week prior to the normal data collection period. Studies in animals chronically-exposed to JP8 have shown selective cellular damage to cells located in the retina and cerebellum. Since any retinal changes detectable by the ERG would be the result of chronic exposure, repeat ERG testing (pre and post work period) was unnecessary. Approximately 20 subjects at each base were selected to complete the ERG.

Exposure measures:

The JP8 exposure measures conducted as part of the JP8 study are briefly discussed below. A more detailed explanation of each exposure measure is provided in the abstracts included in this report.

Biological measures:

Blood: Each subject submitted two 40-ml blood specimens—one specimen during the morning test period and one in the afternoon. Trained phlebotomists from the Air Force Research Laboratory (AFRL) at Wright-Patterson AFB collected all blood specimens. Each blood specimen was divided into three aliquots. Texas Tech University conducted quantitative analysis for the enzyme Glutathione-S-Tranferase (G-S-T) in blood. Researchers from Brooks AFB, in collaboration with researchers from the EPA, conducted analysis for JP8 markers. Scientists from the University of North Carolina analyzed blood specimens for metabolites of benzene and naphthalene. NIOSH and Navy collaborating scientists conducted additional biomarker analyses. A small amount of residual blood from each subject was provided to AFRL for physiologically-based pharmacokinetic (PBPK) modeling of jet fuel metabolism.

Urine: Urine samples collected prior to and after the sampling period were divided into two aliquots. Researchers from the University of North Carolina analyzed urine for the presence of metabolites for benzene and naphthalene. NIOSH conducted analysis of urine samples for the presence of renal biomarkers of exposure.

Breath: Three breath samples were typically collected before and after the work period. Using devices called SUMMA canisters, a scientist from the EPA collected breath samples from selected enrollees and analyzed the specimens for the presence of JP8 markers. Breathe samples, collected using a 75-ml glass bulb collection device, were processed by University of North Carolina scientists to identify the presence of benzene and naphthalene. Using a third breath collection method, a researcher from Johns Hopkins University obtained pre- and post-work samples from selected enrollees and performed an analysis to quantify the amount of JP8 constituents contained in each specimen.

Skin Exposure Sampling: Epidermal specimens were collected prior to and following the work period using a dermal taping method. The skin specimens were analyzed for the presence of naphthalene by researchers from the University of North Carolina.

Body Temperature Monitoring: Internal body temperature, a potential confounding variable in the association between jet fuel constituent metabolism and performance/health measures, was monitored during the enrollee's work period. Selected subjects were asked to swallow a small pill-like sensor. The device provided continuous monitoring of body core temperature during the enrollee's work period. Other enrollees were asked to wear an aural or skin temperature probe. All enrollees wore Polar Band heart rate monitors around the chest area, and activity sensors on the wrist.

Performance/Health Measures:

Enrollees were asked to submit to a series of performance and health effects measures. Tests included the Global Assessment System for Humans/ Behavioral Assessment and Research System (GASH/BARS), the Postural Sway Test, the Eye Blink Conditioned Response Test and the Electroretinogram (ERG). Subjects were also asked complete an electronically administered questionnaire. Medical records were reviewed for pertinent health events occurring during the preceding year.

Global Assessment System for Humans (GASH)/ Behavioral Assessment and Research System (BARS): The GASH/BARS system consists of a series of computer-based neurobehavioral tests designed to measure motivation, response speed, coordination, grip strength, complex mental functioning, memory, and attention. Subjects completed the GASH/BARS test series prior to and after the work period. Data from the subjects' Air Force Qualifying Test (AFQT) were also obtained from the Air Force Personnel Center and used to support the GASH/BARS analysis. AFQT exam scores were coded using subject codes to protect subject confidentiality and ensure study blinding.

Electroretinogram (ERG): The ERG is a device designed to measure the electrical response of the eye to brief, high intensity flashes. In this study, the ERG was used to determine the association between JP8 exposure and retinal Mueller cell function. Subjects who volunteered for this protocol underwent an ERG as part of their evaluation. Since the hypothesized retinal changes are associated with chronic JP8 exposure, the ERG procedure was accomplished only once on the enrollee selected. In addition to comparisons between the exposed and unexposed groups, ERG results were compared with normative data.

Postural Sway: A team of researchers from the University of Cincinnati conducted a series of tests to assess the enrollee's balance. During the test, subjects were asked to perform a series of procedures while standing on a platform designed to measure changes in balance. The procedures included standing on the platform alone and with a foam pad between the platform and enrollee's feet while performing a series of procedures with their eyes open or closed. Each enrollee also answered a short list of questions prior to postural sway testing.

Eye Blink Conditioned Response (ECR): The eye blink response is a reflex that can be classically conditioned. The ECR is considered a sensitive measure of more global issues of brain functioning, and is appropriate for assessing robust and/or subtle changes in neural processing that one might expect from repeated exposure to jet fuel vapors. Enrollees completed the ECR during pre- and post-work periods. Navy technicians conducted this procedure on selected enrollees. The Navy Neurotoxicology Group at Wright-Patterson AFB analyzed the results of ECR tests.

Risk Factor Questionnaire. Each subject volunteer completed a series of questions designed to assess self-reported symptoms, and exposure to potentially confounding factors, such as alcohol and tobacco. Questions regarding hobbies and work-shift history were also addressed. Subjects also completed a series of standardized questions from a copyrighted questionnaire termed the SF-36. Both questionnaires were administered electronically after completion of the GASH/BARS.

Medical Records Review: Epidemiologists from Texas Tech University and AFIERA reviewed the medical records of those enrolled in the study. The epidemiologists recorded health care events occurring during the year prior to the study period using broad disease categories. Associations between health care event frequency and JP8 exposure were tested using these data.

Analysis

Data collection:

Members of the JP8 research team collected all exposure specimens and outcome data. For the most part, the researchers associated with specific sub-protocols (such as the Postural Sway Test) included in the overall JP8 study were responsible for applying tests, analyzing specimens and collecting data specific to their sub-protocols. The exceptions to this rule were time-series logs, study eligibility, exertion and daily activity questionnaires, and blood specimen collection. The blood specimens were collected by AFRL phlebotomists, divided into aliquots and provided to other researchers. The questionnaires and logs were collected by AFIERA personnel, entered into spreadsheets and provided to all researchers to aid in their analyses.

In total, 339 Air Force active duty members were enrolled in the study. Of those enrolled, 324 completed all required tests and submitted all required specimens. Some enrollees were not able to complete the entire study due to unavoidable circumstances. Eight enrollees completed only the ERG, 3 only completed a questionnaire, 2 completed only the ERG and a questionnaire, and 2 enrollees completed all but one or two of the required tests.

Initial Exposure Classification:

Using information provided by the enrollees and exposure stratification assignment at the time of enrollment, those enrolled in the study were categorized into one of four groups. The exposure categories were based on the probability of JP8 exposure in the completion of normal operational duties. The exposure categories and decision tree for enrollee categorization are listed below.

The categorization scheme was developed and employed for two reasons. First, as pointed out in the abstracts following this section, processing specimens is time-consuming. Six months after data collection, some exposure data, particularly measures of JP8 in blood, remained unavailable to researchers. Measures of JP8 effects, particularly neurological test results, were however available for analysis shortly after the end of the data collection phase. The categorization scheme was employed to allow those measuring JP8 effects to obtain a preliminary assessment of degree to which JP8 impacted human performance. The second reason for employing the categorization scheme relates to chronic measure of effects. Effects measures, such as the ERG, medical visit history and self-reported symptoms, are not necessarily related to acute JP8 exposure, but may be influenced by chronic exposure to jet fuel.

CATEGORY	**CRITERIA**		
HI	Classified as exposed by study bioenvironmental engineer? YES		
	AND BY SELF REPORT		
	Does your current job routinely bring you into physical contact with jet fuel? YES		
	What is your primary AFSC? 2A6X4		
	Does working in your primary AFSC bring you into physical contact with jet fuel? YES		
	Age under 35? YES		
HI MOD	Classified as exposed by study bioenvironmental engineer? YES		
	BY SELF REPORT		
	Does your current job routinely bring you into physical contact with jet fuel? YES		
	Does working in your primary AFSC bring you into physical contact with jet fuel? YES		
	AND		
	What is your primary AFSC?		2FOXX 2T3XX 2E4XX
	OR		
	Classified as exposed by study bioenvironmental engineer? YES		
	BY SELF REPORT		
	Does your current job routinely bring you into physical contact with jet fuel? YES/NO		
	Does working in your primary AFSC bring you into physical contact with jet fuel? YES/NO		
	MUST ANSWER NO TO ONE OF THE ABOVE QUESTIONS		
	AND		
	What is your primary AFSC?		2A6X4
	OR		
	Primary AFSC = 2A6X4 and Age 35+		
MOD	Classified as exposed by study bioenvironmental engineer? NO		
	BY SELF REPORT		
	Does your current job routinely bring you into physical contact with jet fuel? YES/ NO		
	Does working in your primary AFSC bring you into physical contact with jet fuel? YES/NO		
	MUST ANSWER YES TO ONE OF THE ABOVE		
	AND		
	What is your primary AFSC?	ANY BUT	2A6X4 2FOXX 2T3XX 2E4XX
LOW	Classified as exposed by study bioenvironmental engineer? NO		
	BY SELF REPORT		
	Does your current job routinely bring you into physical contact with jet fuel? NO		
	Does working in your primary AFSC bring you into physical contact with jet fuel? NO		
	AND		
	What is your primary AFSC?	ANY BUT	2A6X4 2FOXX 2T3XX 2E4XX

Since the categorization scheme assesses the probability of occupational JP8 exposure, it serves well as a means for stratifying enrollees based on chronic exposure.

While the categorization scheme provides a method for determining JP8 exposure, body burden measures employed in this study provide a direct method for quantifying exposure. Direct body burden measures should be far superior to categorization in assessing acute exposure. Three strategies were employed to hasten the availability of JP8 body burden measures.

Strategy One. Prioritization of Specimen Processing

While the processing of all samples is needed to achieve sufficient sample size to determine statistically significant differences in health and performance effects based on JP8 body burden, investigators could estimate the strength of JP8 effects by conducting analyses using a representative sample of the data. To provide this sample as quickly as possible, a stratified random sample of the study enrollees was selected and all investigators were notified to process specimens from these enrollees first. A weighted sample consisted of 110 enrollees chosen at random after stratifying all enrollees by exposure category. Research may refer to this prioritized sample in the abstracts that follow.

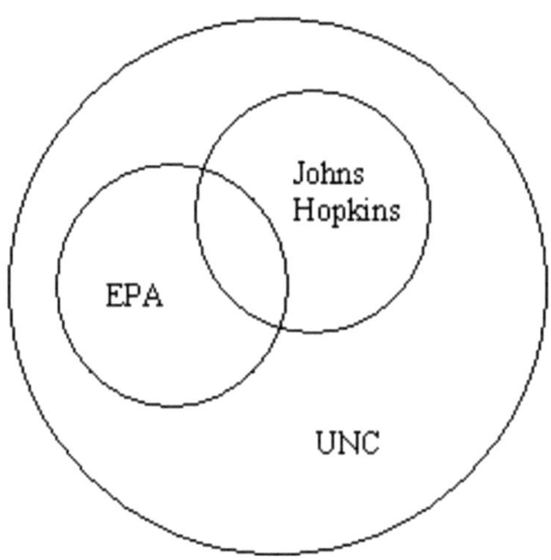

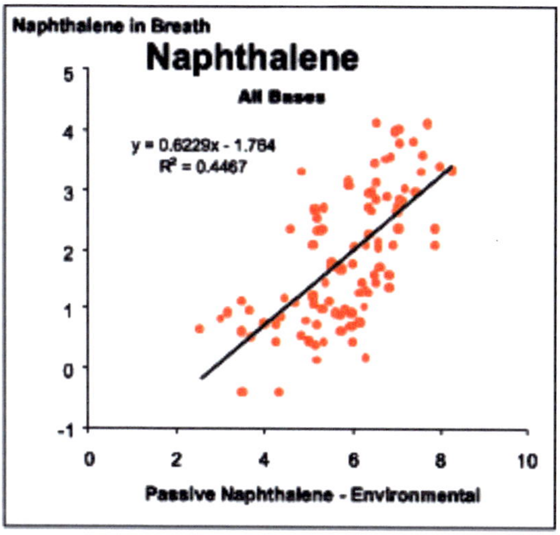

Source: Egeghy, University of North Carolina

Strategy Two. Selection of a Single Continuous Measure of Exposure

In January 2001, a group of JP8 researchers, including the three researchers responsible for measuring JP8 in breath specimens, met to consider whether one single breath measure could sufficiently provide a continuous measure of JP8 body burden for use by other researchers involved in the study. The group reviewed available breath data and found reasonable agreement between the three breath measures with correlation coefficients in the 0.7 to 0.8 range. As graphically represented below, breath measures via EPA and John Hopkins methodologies though highly correlated, were not available for all subjects. The breath measures provided by the University of North Carolina provided data on two constituents of JP8: benzene and naphthalene. Though not as highly correlated as the EPA and Johns Hopkins data, the naphthalene data was generally in agreement with the other breath measures and was available on nearly all enrollees. The exposure measurement team of researchers agreed to provide naphalene measurement data to all collaborating researchers for use in assessing the impact of JP8 acute exposure. The abstracts that follow will, in some cases, refer to these measures in their analyses.

Strategy Three.

Preliminary analysis of JP8 exposure using post-workday naphthalene breath samples failed to demonstrate any association between exposure and

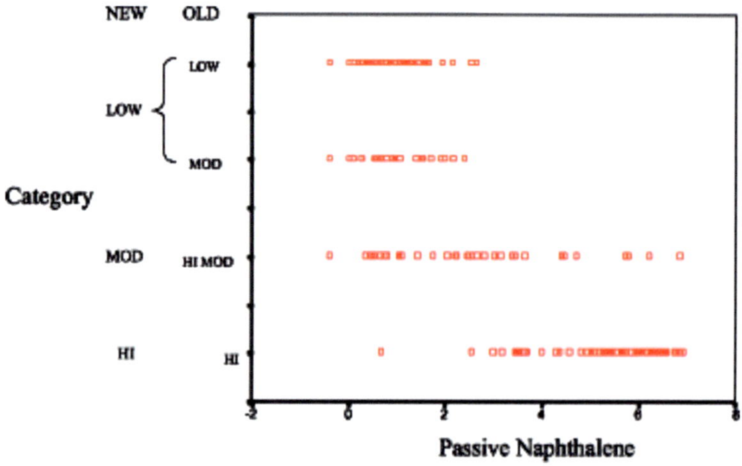

the three primary neurologic tests used in the study. Not only was no association demonstrated, the findings noted during analysis using the exposure categories described under Strategy One were lost. Further, an analysis performed by Egeghy and graphically displayed below showed weak correlations (R2 = -0.44) between post-workday naphthalene specimens and subject-specific environmental naphthalene samples.

Based on these findings, the research team investigated the correlation between environmental naphthalene samples and the previously developed exposure categories. Initial analysis showed much higher correlations (R2 = -0.83). Investigation of outliers revealed the highest environmental naphthalene measures among the LOW exposure category involved subjects tested on Monday and Tuesday at Davis Monthan AFB. Study notes showed investigators were concerned about the possibility of secondary exposure among LOW exposed subjects during post-sampling periods on Monday and Tuesday. LOW exposed subjects returned to the sample collection site at the same time as HI exposed subjects. Further, because of the building design, a strong fuel odor was reported to the research team. Measures were taken to prevent secondary exposure during the following days at Davis Monthan AFB and at subsequent study locations. Based on these findings, the LOW exposed subjects from Monday and Tuesday at Davis Monthan AFB were eliminated from the preliminary analyses.

Outlier analysis also revealed one subject LAN9356 was miscoded. The subject, originally coded as moderately exposed (MOD), actually worked in Aircraft Fuel System Repair Shop and handled fuel-soaked fire suppression foam on the day of the test. After the subject's exposure code was changed and the Davis Monthan ABF LOW-exposed subjects from Monday and Tuesday were removed from the dataset, correlation coefficients improved to R2 = -0.86. Since the LOW and MOD categories were indistinguishable with respect to environmental naphthalene measures, the categories were collapsed into one category term LOW. The HI MOD category was renamed MOD and the HI category remained unchanged. Researchers may refer to this new categorization scheme in the abstracts included in the report.

Risk Characterization:

The full characterization of JP8 health risks, to be conducted by Texas Tech University, will take into account exposure measures of JP8 and measures of effect. The risk characterization will attempt to determine the association between the various measures of effect used in this study and assess the overall impact, by JP8 dose, on workers exposed to the fuel. Details regarding the JP8 risk analysis are provided at the conclusion of this report.

General Results

Enrollment Results

Potential subjects for the Risk Assessment of Acute Exposure to Jet Fuel study were solicited from six Air Force bases in the continental United States. Of the approximately 450 candidates who responded to recruitment efforts, 394 received the study briefing and completed the exclusion criteria questionnaire. From these candidates, 339 subjects met minimal enrollment criteria and entered the study. Enrollment percentages by study site are listed below.

Of the 284 males and 55 females that began the study, 15 withdrew before the study was completed. Eight completed only the electroretinograph (ERG) test, 3 completed only the Risk Factor Questionnaire (RFQ), and 2 completed both the ERG and RFQ before dropping out. Two additional subjects withdrew with after completing nearly all parts of the study. Most of the withdrawals (7) were due to a tropical storm that arrived at Hurlbert Field the week of the study. The storm forced researchers to cancel the final data collection day of the study.

Details regarding the enrolled subjects are included in the table below. A total of 284 men and 55 women were enrolled. Subject ages ranged from 18 to 44 years, with an average age of 26.1 and a median age of 24 years. The exposure categories used in the table below represent the revised categorization discussed in the General Methods section. The LOW categories include subjects with no or rare exposure to JP8. Subjects in the MOD category do not have daily exposure to JP8, but may periodically perform tasks requiring fuel

Study Site	Number Considered	Number Enrolled	Percent Enrolled
Davis Monthan	74	65	87.8
Seymour Johnson	70	49	70.0
Langley	66	59	89.4
Pope	56	60	93.3
Little Rock	64	49	76.6
Hurlbert Field	64	57	89.1
Total	394	339	86.0

Reasons for ineligibility included:

Preexisting medical condition	20
Contraindicated prescription drugs or over-the-counter vitamins	17
Recent surgery	3
Not enough time-on-station	3
TDY during the week of the study	4
On quarters or profile	2
Pregnant	2
Candidate opted out	4
TOTAL	**55**

exposure. The HI category includes only personnel assigned to Aircraft Fuels System Maintenance shops.

Exposure categories were reasonably comparable with respect to right-handedness, race, height, weight and body mass index. The percentage of smokers, alcohol, caffeine, and processed meat users is approximately the same in all categories. Subjects report approximately the same number of hours worked in a week and engage in same amount of physical activity off duty. While subjects in the highest JP8 exposure categories were, on average, younger than those in the other exposure categories, the biological significance of these age differences is questionable. No differences are seen in the months on the base and in the current job. In two different physical exertion measures, however, the amount of physical work required to perform duties associated with their job, is significantly greater (P-value < 0.001) for those in the HI category. No differences are seen in the mental exertion required. Male sub-

Risk Assessment of Acute Exposure to Jet Fuel
Personal Characteristics

	Exposure Status					
	Males = 284			Females = 55		
	HI	MOD	LOW	HI	MOD	LOW
Age (mean)	24.6	26.8	27.6	22.6	33.8	24.8
Age by group (%)						
Under 20	3.5	2.6	3.1	20.0	0.0	5.0
20 to 24	57.4	48.7	42.3	50.0	0.0	55.0
25 to 29	27.8	20.5	20.0	20.0	20.0	25.0
30 to 34	4.3	7.7	15.4	0.0	20.0	7.5
35 to 39	5.2	17.9	13.8	0.0	60.0	5.0
40 and over	1.7	2.6	5.4	0.0	0.0	2.5
Right Handed (%)	83.5	76.9	83.8	100.0	80.0	95.0
Caucasian (%)	80.0	74.4	72.8	80.0	100.0	70.0
Height in inches (mean)	70.5	70.5	70.7	66.2	66.2	65.0
Weight in lb (mean)	178.0	182.3	186.7	143.7	151.4	145.0
Body mass index (mean)	28.4	27.7	27.2	31	29.7	29.5
Smoke at least 1/4 pack per day (%)	43.9	47.4	32.3	50.0	60.0	30.8
Alcohol users (%)	61.7	66.7	74.2	75.0	100.0	64.1
Daily caffeine users (%)	51.9	66.7	53.5	75.0	60.0	28.2
Eat processed meats >1 time per week (%)	43.5	35.9	40.3	37.5	20.0	28.9
Months of the job	53.8	49.2	57.5	31.9	34.6	30.55
Months on current base	33.3	28.5	33.2	19.8	43.2	26.9
Engage in physical activity 1-2 times per week (%)	62.5	66.7	60.7	100.0	100.0	61.1
Work 8-10 hours per day at job (%)	94.4	84.6	85.9	75.0	100.0	89.7
A great deal of physical work is required as part of job (%)	13.9	5.1	7.0	25.0	0.0	0.0
Physical exertion score (mean)	10.7	9.3	7.0	10.2	8.4	5.6
Mental exertion score (mean)	3.9	3.5	4.0	3.5	4.0	3.9
Find job very stressful (%)	26.2	5.3	19.5	12.5	20.0	5.1

jects in the HI category are more likely to report a great deal of stress associated with their job. Due to much smaller sample size, comparisons of personal characteristics among females are less stable than those of males.

JP8 exposure results and the impact of JP8 exposure on various performance tests and health outcome measures are reported in the abstract found in this report.

Appendix C

Review of Tests Assessing Neurologic Function in Persons Exposed to Jet Fuels

This appendix described four types of tests that have been used to assess the neurotoxicity potential of jet-propulsion fuel 8 and related fuels in humans: clinical neurological examinations, posturograms, nerve conduction studies, vibration sensation studies, and blink reflex classical conditioning studies. For each test, a critical analysis of its use in assessing neurotoxicity from exposure to jet fuels is presented. Limitations of these tests are also presented.

CLINICAL NEUROLOGIC EXAMINATIONS

Neurological examinations of 29 aircraft factory workers chronically exposed to jet fuel vapors revealed findings consistent with central and peripheral nervous system involvement (Knave et al. 1976). The exposed subjects were classified into two groups: heavily exposed (n = 13) and less heavily exposed (n = 16). Subjects from both groups (all of the heavily exposed workers and 7 of the 16 less heavily exposed workers) reported having repeatedly experienced acute effects of jet fuel exposure such as dizziness, headache, nausea, respiratory tract symptoms, heart palpitations, and a feeling of pressure on the chest. Symptoms indicative of peripheral neuropathy included muscle cramps, distal paresthesias, numbness, and paresis. Signs indicative of peripheral neuropathy on clinical neurological examination included reduced sensory perception particularly of pain and temperature sense among the subjects from the heavily exposed group. A high rate of symptoms indicative of CNS involvement (neurasthenia and psychasthenia) was also observed in the two exposed groups on comparison with the reference groups.

POSTUROGRAM

Posturogram measures spontaneous body sway, which is considered to be a quantitative version of the Romberg test, which is a simple clinical test of the integrated postural pathways and mechanisms (i.e., peripheral nerve fibers, spinal cord pathways, vestibular system components, and the cerebellum). The posturogram is a way of measuring a person's ability to maintain an upright posture against variable perturbations. Performance on posturogram is a function of gender, age, and vision (Kollegger et al. 1992; Black et al. 1982; Thyssen et al. 1982). Various posturography techniques are available to provide quantitative measures. Posturogram performance deficits have been associated with exposures to various solvents including toluene, xylene, and jet fuel (Smith et al. 1997; Yokoyama et al. 1997).

The validity and reliability of the posturogram as a measure of central and peripheral nervous system function has been demonstrated in various studies. Posturogram findings have been shown to positively correlate with other measures of vestibular function such as positional nystagmus demonstrating the validity of this test (Kubo et al. 1990). Studies in subjects who consumed ethanol immediately before testing showed disturbances in body sway that correlated with the amount of ethanol consumed indicating that this test is a sensitive and valid measure of CNS dysfunction induced by neurotoxicants that are known to affect balance and the severity of the deficit can reflect the exposure dose (Mills and Bisgrove 1983; Lukas et al. 1989; Kubo et al. 1989). Studies by Benvenuti et al. (1999) demonstrated the consistency of posturogram findings on test-retest among geriatric subjects with disequilibrium demonstrating the reliability of this test as well as its validity.

Although the results of a study by Uimonen et al. (1995) indicate that the posturogram of a malingerer can be differentiated from that of a subject with vestibular neuritis by body sway velocity, these findings should be interpreted with caution. The administration of neuropsychological tests sensitive to malingering and motivation such as the Test of Memory and Motivation (TOMM) can be administered along with the posturogram and other conventional tests of neurological function to further differentiate those patients with a true performance deficit from those with deficits induced by factors such as secondary gain. Patients involved in litigation who have abnormal TOMM scores and abnormal posturograms may have motivational influences that are affecting their performances.

Unfortunately, although the sensitivity of posturography is relatively high, the specificity of this test as a measure of neurotoxicant exposure-induced effects is relatively low as it reveals deficits due to many factors (e.g., diabetes and head injury). Thus, the findings on posturogram must also be interpreted in light of their correlation with other measures of neurophysiological and neuropsychological function as well as with the patient's medical history and his or her history of neurotoxicant exposure to parse out the locus of the lesion (i.e., central v. peripheral). For example, the Digit Symbol Test is a neuropsychological measure of psychomotor function that has been shown to positively correlate with posturography findings among patients exposed to ethanol and central acting pharmaceuticals (Lukas et al. 1989; Allen and Lader 1992). Nerve conduction studies have been shown to positively correlate with posturography findings among patients with diabetic neuropathy (Uccioli et al. 1995).

In conclusion, posturography is a reasonable test to use to corroborate with data from neurophysiological and neuropsychological tests but performance on this test cannot be relied upon exclusively to establish exposure limits for neurotoxicants encountered in the workplace.

NERVE CONDUCTION STUDIES

Peripheral neuropathy has been reported in subjects exposed to various solvents including *n*-hexane and jet fuels. Neurological and neurophysiological examinations of 29 aircraft factory workers chronically exposed to jet fuel vapors findings consistent with peripheral neuropathy (Knave et al. 1976). The exposed subjects were classified into two groups: heavily exposed (n = 13) and less heavily exposed (n = 16). Clinical neurological examinations, nerve conduction velocities studies, and assessments of vibration sensation thresholds were performed on all subjects. All subjects in the heavily exposed group and 7 of 16 from the less heavily exposed reported having repeatedly experienced acute effects such as dizziness, headache, nausea, respiratory tract symptoms, heart palpitations, a feeling of pressure on the chest during exposures to jet fuel vapors in their inhaled air. A high rate of symptoms indicative of neurasthenia and psychasthenia and symptoms and signs indicative of polyneuropathy was observed both in the heavily exposed group and in the two groups combined in comparison with reference groups.

Nerve conduction studies have been used extensively to study peripheral nerve function. These tests have been shown to be valid marker of nerve damage and a correlated with pathological findings. These tests have been shown to be reliable with good retest reliability and good interrator reliability due to established normal values and stringent testing procedures and protocols (Kimura 2001). Based on its review of the literature, the subcommittee concludes that nerve conduction studies are a good marker of neurological injury due to exposures to neurotoxicants and that the findings from this test can be relied upon to establish occupational and environmental exposure limits.

VIBRATION SENSATION

Vibration sensation thresholds were used to assess functioning in 29 aircraft factory workers chronically exposed to jet fuel vapors and who reported symptoms consistent with peripheral neuropathy (Knave et al. 1976). The exposed subjects were classified into two groups: heavily exposed (n = 13) and less heavily exposed (n = 16) and were compared with unexposed controls. Comparison of the subjects from the high-exposure group with controls revealed significant differences on vibration sensation thresholds.

The vibration sensation test is purported to be a quantitative measure of perception of a vibrating stimulus. It is used to see if a person can volitionally tell the examiner when a stimulus is first perceived. It is intended to reveal impairment in the ability of a peripheral nerve to conduct an impulse. If the subject has a high threshold for this test, it is suggested by the amount of time the subject requires for acknowledging his/her perception of the stimulus. This test is subject to embellishment by the examinee and to observer bias by the test administrator. Such impairments detected on screening tests such as this must be correlated with other more objective tests for evidence of peripheral neuropathy such as nerve conduction velocities, tendon reflex responses, and patterns of sensory loss detected by pin prick perception.

In conclusion, the subcommittee's review of the literature suggests that the use of this test to assess peripheral nerve function among persons exposed to neurotoxicants is reasonable if corroborated with data from conventional validated neurophysiological tests. However, performance on this test cannot be relied upon by itself to establish exposure levels for neurotoxicants.

BLINK REFLEX CLASSICAL CONDITIONING

The use of blink reflex classical conditioning to investigate motor learning in subjects exposed to neurotoxicants has been suggested. Bekkedal et al. (2001) reported that the blink reflex conditioning response may be affected by exposure to JP-8. It has been shown that the cerebellum is involved in the acquisition of motor skills and procedural learning (Corkin 1968; Sanes et al. 1990; Laforce and Doyon 2001). The data reviewed indicate that the blink reflex classical conditioning test is a valid marker of cerebellar function (Bracha et al. 1997; Glocker et al. 1999; Sears et al. 2000; Sommer et al. 2001). However, there are few comprehensive studies supporting the validity of this test by showing that it correlates positively with other measures of motor function and does not correlate positively with irrelevant measures (Sommer et al. 2001). Furthermore, the interrator reliability and test-retest reliability have not been fully established for this test. In addition, it has not been shown that this test would be practical to use from both a cost and risk versus benefits perspective in research studies.

The subcommittee concludes that the use of this test to assess cerebellar function among persons exposed to neurotoxicants is reasonable, but the data should not be relied upon exclusively and must be corroborated with data from conventional validated neuropsychological and neurophysiological tests particularly if it is to be used to establish exposure levels for neurotoxicants.

REFERENCES

Allen, D., and M. Lader. 1992. The interactions of ethanol with single and repeated doses of suriclone and diazepam on physiological and psychomotor functions in normal subjects. Eur. J. Clin. Pharmacol. 42(5):499-505.

Bekkedal, M.Y.V., S.M. McInturf, G.D. Ritchie, and J. Rossi III. 2001. Eyeblink conditioning response test used to assess performance in JP-8 exposed air force personnel. Pp. 69-71 in JP8 Final Risk Assessment. The Institute of Environmental and Human Health (TIEHH), Lubbock, TX. August 2001.

Benvenuti, F., R. Mecacci, I. Gineprari, S. Bandinelli, E. Benvenuti, L. Ferrucci, A. Baroni, M. Rabuffetti, M. Hallett, J.M. Dambrosia, and S.J. Stanhope. 1999. Kinematic characteristics of standing disequilibrium: Reliability and validity of a posturographic protocol. Arch. Phys. Med. Rehabil. 80(3):278-287.

Black, F.O., C. Wall III, H.E. Rockette Jr., and R. Kitch. 1982. Normal subject postural sway during the Romberg test. Am. J. Otolaryngol. 3(5):309-318.

Bracha, V., L. Zhao, D.A. Wunderlich, S.J. Morrissy, and J.R. Bloedel. 1997. Patients with cerebellar lesions cannot acquire but are able to retain conditioned eyeblink reflexes. Brain 120(Pt 8):1401-1413.

Corkin, S. 1968. Acquisition of motor skill after bilateral medial temporal-lobe excision. Neuropsycologia 6(3):255-265.

Glocker, F.X., M. Lauk, D. Foll, B. Koster, B. Guschlbauer, J. Timmer, G. Deuschl, and C.H. Lucking. 1999. Classical conditioning of the electrically elicited blink reflex in humans: A new method of data analysis. J. Neurosci. Methods 89(2): 133-140.

Kimura, J. 2001. Electrodiagnosis in Diseases of Nerve and Muscle: Principles and Practice, 3rd Ed. New York: Oxford University Press.

Kollegger, H., C. Baumgartner, C. Wober, W. Oder, and L. Deecke. 1992. Spontaneous body sway as a function of sex, age, and vision: Posturographic study in 30 healthy adults. Eur. Neurol. 32(5):253-259.

Knave, B., H.E. Persson, J.M. Goldberg, and P. Westerholm. 1976. Long-term exposure to jet fuel: An investigation on occupationally exposed workers with special reference to the nervous system. Scand. J. Work Environ. Health 2(3):152-164.

Kubo, T., Y. Sakata, T. Matsunaga, A. Koshimune, S. Sakai, K. Ameno, and I. Ijiri. 1989. Analysis of body sway pattern after alcohol ingestion in human subjects. Acta Otolaryngol. 468(Suppl):247-252.

Kubo, T., Y. Sakata, A. Koshimune, S. Sakai, K. Ameno, and I. Ijiri. 1990. Positional nystagmus and body sway after alcohol ingestion. Am. J. Otolaryngol. 11(6):416-419.

Laforce Jr., R., and J. Doyon. 2001. Distinct contribution of the striatum and cerebellum to motor learning. Brain Cogn. 45(2):189-211.

Lukas, S.E., B.W. Lex, J.P. Slater, N.E. Greenwald, and J.H. Mendelson. 1989. A microanalysis of ethanol-induced disruption of body sway and psychomotor performance in women. Psychopharmacology (Berl) 98(2):169-175.

Mills, K.C., and E.Z. Bisgrove. 1983. Body sway and divided attention performance under the influence of alcohol: Dose-response differences between males and females. Alcohol Clin. Exp. Res. 7(4):393-397.

Sanes, J.N., B. Dimitrov, and M. Hallett. 1990. Motor learning in patients with cerebellar dysfunction. Brain 113(Pt 1):103-120.

Sears, L.L., N.C. Andreasen, and D.S. O'Leary. 2000. Cerebellar functional abnormalities in schizophrenia are suggested by classical eyeblink conditioning. Biol. Psychiatry 48(3):204-209.

Smith, L.B., A. Bhattacharya, G. Lemasters, P. Succop, E. Puhala II, M. Medvedovic, and J. Joyce. 1997. Effect of chronic low-level exposure to jet fuel on postural balance of US Air Force personnel. J. Occup. Environ. Med. 39(7):623-632.

Sommer, M., J. Grafman, I. Litvan, and M. Hallett. 2001. Impairment of eyeblink classical conditioning in progressive supranuclear palsy. Mov. Disord. 16(2):240-251.

Thyssen, H.H., J. Brynskov, E.C. Jansen, and J. Munster-Swendsen. 1982. Normal ranges and reproducibility for the quantitative Romberg's test. Acta Neurol. Scand 66(1):100-104.

Uccioli, L., P.G. Giacomini, G. Monticone, A. Magrini, L. Durola, E. Bruno, L. Parisi, S. Di Girolamo, and G. Menzinger. 1995. Body sway in diabetic neuropathy. Diabetes Care 18(3):339-344.

Uimonen, S., K. Laitakari, H. Kiukaanniemi, and M. Sorri. 1995. Does posturography differentiate malingerers from vertiginous patients? J. Vestib. Res. 5(2):117-124.

Yokoyama, K., S. Araki, K. Murata, M. Nishikitani, K. Nakaaki, J. Yokota, A. Ito, and E. Sakata. 1997. Postural sway frequency analysis in workers exposed to n-hexane, xylene, and toluene: Assessment of subclinical cerebellar dysfunction. Environ Res. 74(2):110-115